Advances in Geographical and Environmental Sciences

Series Editor

R. B. Singh, University of Delhi, Delhi, India

Advances in Geographical and Environmental Sciences synthesizes series diagnostigation and prognostication of earth environment, incorporating challenging interactive areas within ecological envelope of geosphere, biosphere, hydrosphere, atmosphere and cryosphere. It deals with land use land cover change (LUCC), urbanization, energy flux, land-ocean fluxes, climate, food security, ecohydrology, biodiversity, natural hazards and disasters, human health and their mutual interaction and feedback mechanism in order to contribute towards sustainable future. The geosciences methods range from traditional field techniques and conventional data collection, use of remote sensing and geographical information system, computer aided technique to advance geostatistical and dynamic modeling.

The series integrate past, present and future of geospheric attributes incorporating biophysical and human dimensions in spatio-temporal perspectives. The geosciences, encompassing land-ocean-atmosphere interaction is considered as a vital component in the context of environmental issues, especially in observation and prediction of air and water pollution, global warming and urban heat islands. It is important to communicate the advances in geosciences to increase resilience of society through capacity building for mitigating the impact of natural hazards and disasters. Sustainability of human society depends strongly on the earth environment, and thus the development of geosciences is critical for a better understanding of our living environment, and its sustainable development.

Geoscience also has the responsibility to not confine itself to addressing current problems but it is also developing a framework to address future issues. In order to build a 'Future Earth Model' for understanding and predicting the functioning of the whole climatic system, collaboration of experts in the traditional earth disciplines as well as in ecology, information technology, instrumentation and complex system is essential, through initiatives from human geoscientists. Thus human geosceince is emerging as key policy science for contributing towards sustainability/survivality science together with future earth initiative.

Advances in Geographical and Environmental Sciences series publishes books that contain novel approaches in tackling issues of human geoscience in its broadest sense — books in the series should focus on true progress in a particular area or region. The series includes monographs and edited volumes without any limitations in the page numbers.

More information about this series at http://www.springer.com/series/13113

Asheem Srivastav

Energy Dynamics and Climate Mitigation

An Indian Perspective

 Springer

Asheem Srivastav
Gandhinagar, Gujarat, India

ISSN 2198-3542   ISSN 2198-3550   (electronic)
Advances in Geographical and Environmental Sciences
ISBN 978-981-15-8939-3   ISBN 978-981-15-8940-9   (eBook)
https://doi.org/10.1007/978-981-15-8940-9

This Springer imprint is published by the registered company Springer Nature Singapore Pte Ltd.
The registered company address is: 152 Beach Road, #21-01/04 Gateway East, Singapore 189721,
Singapore

Contents

Chapter 1
Energy Security and Sustainability: An Overview

Abstract Every living entity requires internal energy, derived through biological process, for survival, growth, and sustenance. Human being is an exception in the sense that they need additional energy from external sources to accomplish their quest for excellence. The availability and adequacy of external energy and reliability of its supply whether commercial or non-commercial affect economic productivity, development, and sustenance. The process of global split between energy-rich and energy-poor that started with industrial revolution continues till date and each nation is working relentlessly to ensure secure supply of energy from renewable and non-renewable sources. Energy indicators are developed, and projections are done to assess future energy scenario and an understanding of complete energy flow and bottlenecks. India has been undergoing transformative changes since independence with population, poverty, and energy as the pivotal challenge. With the passage of every day, India faces the toughest challenge of energizing its economic development through power generation that is highly dependent on imports of oil, gas, solar panel, and wind turbines. With climate-related disasters becoming too intense and frequent, India has no option but to race ahead in replacing dirty fuels with renewables.

Keywords Calorie · Joule · Watt · Tons of oil equivalent · BTU · Tons of coal equivalent · Energy security · Energy access · Calorific value · British ton · American ton · Metric ton · Energy indicators

The Quest for Energy: From Carbohydrate to Hydrocarbon and beyond (An illustration)

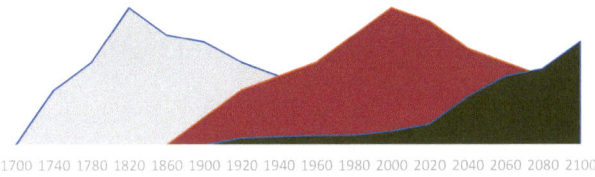

© Springer Nature Singapore Pte Ltd. 2021
A. Srivastav, *Energy Dynamics and Climate Mitigation*,
Advances in Geographical and Environmental Sciences,
https://doi.org/10.1007/978-981-15-8940-9_1

The concept of energy security is of recent origin and has emerged due to high demand for commercial energy consumed for manufacturing goods, providing various services, and for human comfort. Human quest for excellence in knowledge combined with hard work, innovation, perseverance, and dedication has changed the global energy landscape from carbohydrate (wood-based) to hydrocarbon (coal, oil, and gas) economy. Unfortunately, the side effects of hydrocarbon economy have led to many catastrophic phenomena, including ozone hole, pollution of terrestrial and aquatic systems, ocean acidification, and global warming. The industrial revolution that has withstood the test of time so far with external energy derived from non-biological sources has failed the test of sustainability. Wood dominated the socioeconomic development in Europe until the 1500s and was the most important material for building construction, ships, for cooking and heating, as well as for purification of various metals. Rampant and excess extraction of wood eventually led to the disappearance of natural forests in early 1700s. Disappearance of wood was succeeded by emergence of coal in England. Coal was initially obtained through surface mining to avoid risking human lives due to flooding of underground mines. The desire for more coal culminated in invention of steam engine by Thomas Newcomen that saved Britain from energy crisis. By 1800, Britain switched over from carbohydrate to hydrocarbon economy. Hydrocarbon economy has been meticulously controlled by a few nations who have vast resources of coal and oil. These conglomerates have manipulated international price to their advantage frequently threatening the energy security of the world. To counter these forces, many countries, started working on alternative sources of energy such as nuclear (fission in particular), solar, wind, geothermal, and biofuel. In recent years, the climate change disasters have struck humanity with impunity and the scientific community is united in its opinion to stop the use of coal and oil in near future. As the fourth industrial revolution progresses, the world will witness emergence of renewable energy from different sources—and eventually bid farewell to hydrocarbon economy.

1.1 Understanding Various Terms and Conversion Factors

Before embarking on the contents of this chapter, it will be useful for many, if not most, readers to understand the various terms and units of measurement adopted by different countries/institutions/authors.

There are six most common units of energy used at global level (Foresti et al. 2010):

- Calorie
- Joule
- Watt
- BTU (British thermal unit)
- TOE (Tons of oil equivalent); and
- TCE (Tons of coal equivalent)

Almost every science student is conversant with the term 'calorie' and 'joule'. The term *'calorie'* indicates the quantum of energy required to raise the temperature of 1 g of water by one degree centigrade or Celsius (i.e. from 14.5 to 15.5 °C) (Parr 2011), whereas *'joule'* is a measure of quantum of energy (or work) required to generate one watt of power for one second (Chen et al. 2017). '*Watt*', on the other hand, is used to determine power and is equivalent to the rate at which one ampere of current flows through a circuit with potential difference of one volt. Another difference between '*Joule and Watt*' is that the former is a unit of *energy* and the latter is a unit of *energy transfer* in joule per second. In other words, 'Joule' is the amount of energy and 'Watt' measures the rate at which energy is used. For example, when a light bulb (sold in Indian market as 10, 20, 40, 60, 100 watts) of 100 watts glows for 10 seconds, it will consume $100 \times 10 = 1000$ Joules of energy.

We can also express the relationship between joule and watt as follows:

i.   Joule = Watt × seconds
ii. Watt = Joule/second

In many scientific as well as non-science literature, the words 'energy' and 'power' (Box 1) are interchangeably used. For example, we use the term thermal power or nuclear power plant/s but not thermal energy plant or nuclear energy plant. Whereas at the policy level, we invariably use the word 'energy'. Nevertheless, there is a clear distinction between these two words. Power means the rate at which energy is either generated or consumed in a power station and is measured in watts (or kilowatts or megawatts or gigawatt or terrawatt), which in turn means energy per unit time. For example, all power stations have a rating in terms of megawatt or gigawatt. This rating indicates the maximum power output a station can achieve at a given point in time. For example, 1000 gigawatt or 5000 megawatt, and so on. On the other hand, the annual energy output of a power station is mentioned in terms of megawatt-hour or gigawatt-hour or terrawatt-hour. For example,

Power output of 1 terrawatt-hour per year = 1×10^{12} watt-hour/(365 days × 24 hour per day) = 114 megawatt of constant power output for one year.

Box-1 ENERGY AND POWER

Energy is defined as the ability to do work and is measured in joules (J). One joule is the work done when a force of one newton (N) is applied through a distance of one meter. A newton is the unit of force that, while acting on a mass of one kilogram, increases its velocity by one meter per second every second along the direction in which it acts.

Power, on the other hand, is the rate at which energy is transferred and is commonly measured in watts (W), where one watt is one joule per second.

Newton, joule, and watt are defined in the International System of Units. The oil industry measures energy as tons of oil equivalent or ToE where 1 ToE = 41.87 x 109 Joules and barrels of oil equivalent or BoE where 1 BoE = 5.71 x 109 Joules. In the same manner the coal industry measures energy as tons of coal equivalent where 1 TCE = 29.31 x 109 Joules. Commercial electricity is measured as kilo watthour where 1 kWh equals 3.6 x 106 Joules.

Source [UNDP, 2000]

Besides, calorie and joule, another unit of energy frequently used at international level is BTU. The acronym stands for British Thermal Unit (Davis and Wood 1974) and is equivalent to the amount of heat required to raise the temperature of one pound of water through 1° Fahrenheit. One BTU is equivalent to 252 cal or 0.252 kcal or 0.293 watt-hours or 1055 joules.

1 BTU = 252 cal = 0.252 kcal = 1055 Joules = 0.293 watt-hours

In Indian context, consumers pay their electricity bill in proportion to the 'units' of energy consumed by them. One unit of energy consumed indicates use of an electrical item (e.g., a bulb) of 1000 watt (or 1 Kilowatt) for 1 hour. In other words, one unit is equivalent to 1 kilowatt-hour of energy consumed. In terms of BTU, 1 kilowatt-hour generates 3412 BTU.

One unit = 1 kilowatt-hour = 3412 BTU = 853,000 cal

At the global level, a normalized unit of energy is being used which is referred to as KgOE (kilograms of oil equivalent) and TOE (tons of oil equivalent) (see Box 2) (Zou 2020). By convention it is equivalent to the approximate amount of energy that can be extracted from 1 kilogram or 1 ton of crude oil, respectively. The World Bank, for example, uses KgOE for energy use per capita in its 'Little Green Data Book' which is published annually.

Box-2 TOE (Tons of Oil Equivalent)	
1 TOE equals	11.63 Mega-watt Hour (MWH)
	14.868 Giga Joules (GJ)
	39,683,207 British Thermal Unit (BTU)
	1.43 Tons of Coal Equivalent (TCE)
	7.33 Barrels of Oil Equivalent (BOE)

Readers frequently encounter two terms, viz., 'ton' and 'tonne'. Globally, three different types of 'Ton' (https://www.quora.com) have been recognized and these are as follows (Box 3):

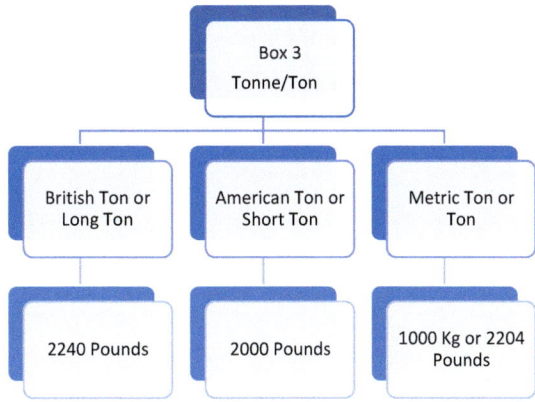

Let us examine the calorific value and quantum of energy (Box 4) that can be generated from some well-known commercial and non-renewable sources (Yearbook, Energy Statistics 2013).

Box 4 Calorific value and quantum of energy			
	Equals BTU (British Thermal Unit)	Total Calorific Value (Calories)	Kilowatt Hours that can be generated
One KWH of Electricity	3412	853000	1.008
One Barrel of Petroleum / crude oil (42 Gallons)	5800000	1461600000	1713.48183
One Barrel of Residual Fuel Oil	627000	158004000	185.233294
One Gallon Gasoline	124238	31307976	36.7033716
One Gallon Diesel	138690	34949880	40.9728957
One Gallon Heating Oil	138690	34949880	40.9728957
One CFT Natural Gas	1023	257796	0.30222274
One Gallon Propane	91333	23015916	26.9823165
One Short (A Short ton is a US term and is equivalent to 2000 pounds or 907 kg, while a Long ton is a British term which is equivalent to 2240 pounds.) ton of Coal	19858000	5004216000	5866.60727
Source [Yearbook, Energy Statistics 2013			

Since different types of fuel materials are used for generating commercial energy, their efficiency (Planning Commission 1953) is compared in terms of calorie content. For example, one ton of crude oil is worth 10 billion cal or 42 billion joules. Similarly, one million ton of Indian coal is worth 4.1 billion cal. It is also possible to measure electrical energy in terms of thermal energy. For example, one billion kilowatt-hour equals 0.86 billion cal. Taking the thermal efficiency of a power plant and other losses in the system, the equivalence between electricity and fossil fuels would be one billion kilowatt-hour = 0.28 million tons of oil equivalent (in case of coal-fired

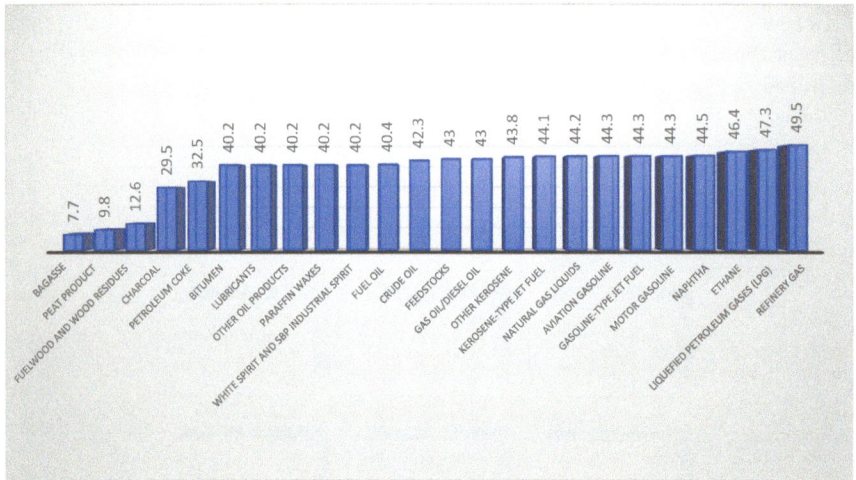

Fig. 1.1 Calorific value of materials used in energy generation (Gigajoule/ton) (Ref-Yearbook, Energy Statistics 2013)

boilers) and 0.261 million tons of oil equivalent (in case of nuclear electricity) (Mason and Mor 2009). One billion kilowatt-hour generated from hydroelectricity or wind power, however, are considered as equivalent to 0.086 million tons of oil equivalent since there is no intermediate stage of heat production while using these primary energies.

A comparison of calorific value of different materials used for energy generation is given in Fig. 1.1.

Bagasse has minimum calorific value (7.7 gigajoule per ton) whereas most of the petroleum derivatives have calorific value in the range of 40–49 gigajoule per ton. Fuelwood too has low calorific value less than a fourth of petroleum derivatives. However, coal generates more energy than petroleum products and byproducts as can be seen in Fig. 1.2.

In many poor countries of Asia and Africa, people rely heavily on thermal energy from firewood, charcoal, plant, and animal residues as well as human and animal energy (Srivastav 2019). Most of these sources are referred to as non-commercial (and non-conventional), although they are often bought and sold in organized markets. Despite their valuable contribution, neither national nor international institutions have yet given sufficient attention to the sources of such non-commercial energy and technologies being used, their economic and environmental consequences, or to the development of alternatives. The acute scarcity of reliable information in this sector calls for more attention to data gathering and research as deforestation and fuelwood shortages have become critical and are appropriately labeled 'the other energy crisis' (Eckholm 1975).

In India, poverty and unprecedented rise in the population has forced a substantial part of the population to continue the use of wood, agriculture waste, and animal waste for generating heat energy to meet the cooking and other heating requirements. Unfortunately, this form of energy continues to be categorized as 'unorganized and

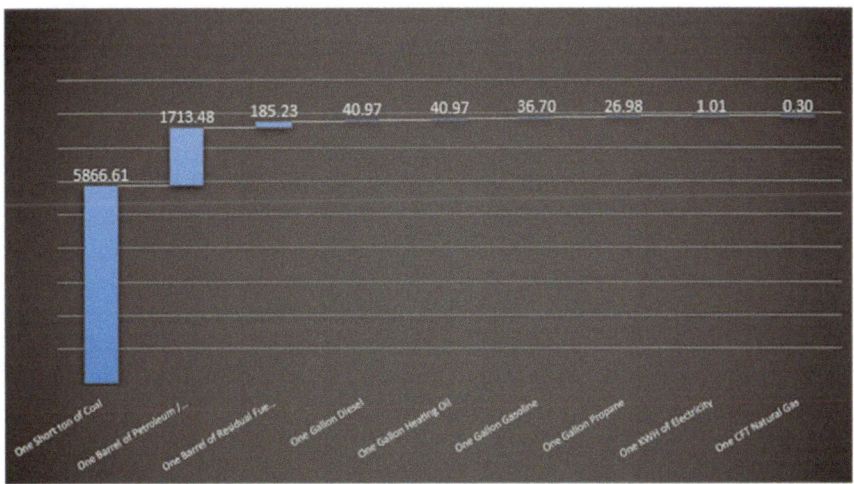

Fig. 1.2 Kilowatt hours generated from different source (Ref-Yearbook, Energy Statistics 2013)

non-commercial' and the planning process does not consider as to how much electricity or gas will be required in future to offset the use of biomass. Every year millions of tons of wood, animal, and plant waste is burnt for cooking food and other purposes. It is a 'no-win' situation as replacing this biomass with coal or gas will cost billions of rupees (for import, extraction, transportation, and distribution). No replacement or slow replacement, on the other hand, would mean loss of trees, soil, and water. For the sake of understanding, Box 5 gives the conversion ratio between wood fuel and coal, crop waste and dung (Srivastav and Srivastav 2015).

> Box 5
>
> 1 Cubic meter of wood (in volume) = 0.725 Metric Ton (in weight)
>
> 1 Metric ton of wood fuel = 0.7 Metric ton of coal
>
> 1 Metric ton of wood fuel = 1.53 Metric ton of crop waste
>
> 1 Metric ton of wood fuel = 2.35 Metric ton of dried animal dung
>
> Source - Srivastav et al., 2015

1.2 Energy Indicators

Different sectors of economy such as residential, commercial, transport, services, and agriculture require energy in different forms. For example, housing may need electricity, gas, and solar water heaters; transport may require petrol, diesel, gas, electricity, and so on. Availability of adequate energy and reliability of supply whether commercial or non-commercial affect productivity, development, and sustenance. It is, therefore, useful to understand various energy indicators to help policy makers and executors in improving industrial and agricultural productivity, job creation, services, and other economic activities in a country. Energy indicators provide a glimpse of energy scenario and an understanding of complete energy flow and bottlenecks. These are indispensable tools for identifying and understanding energy trends so that decision makers can prioritize interventions and consumption pattern. Indicators keep evolving overtime depending on country priorities, requirements, and capabilities. These indicators (Energy Statistics 2018) are primarily economic indicators of energy and are divided into two sub-indicators, viz., use and production pattern and security. Use of energy and production pattern depends on various factors and sub-factors (Fig. 1.3a, b).

The sub-indicators (Energy Statistics 2018) for energy security include fuel stocks and imports and are described in Fig. 1.4.

1.3   The Paradox of Energy Security and Sustainability

Every living entity requires energy for survival, growth, and procreation of itself. However, the case of human being is different from other living creatures in the sense that they need an external source of energy to accomplish their quest for excellence. Societies cannot survive without incessant use and supply/demand of energy and the original source of energy is human energy—the energy generated by human cells for supplying to muscles. As time passed, humans mastered the science of controlling fire and used combustible materials like wood to cook food, and reshaped metals such as iron to make implements. Subsequently, the energy of flowing water and wind was also harnessed, followed by those of animals before we arrived at industrial revolution and started using coal, oil, and gas for generating energy at commercial scale.

To better understand, let us divide energy into internal energy and external energy. Internal energy comes from consumption of food, air, and water that help us in building protein, fat, carbohydrates, and minerals through a long and complex process and storing it in trillions of cells. Each cell contains hundreds/thousands of mito-chondria (biological battery) that contain a substance called adenosine tri phosphate (ATP), which is also known as energy currency of the cell. ATP combines with water to form adenosine di phosphate (ADP) and phosphorus and generates energy within the cell to perform its designated functions. The economics of this conversion process (from ATP to ADP and vice versa) is little understood since this has not been a matter of much concern for the policy makers.

External energy, on the other hand, is required for human comfort, knowledge advancement, exploration, and development in general. This energy is obtained through three processes. One, by direct consumption of combustible materials such as wood, animal waste, coal, charcoal, and kerosene; two, by producing bulk electricity

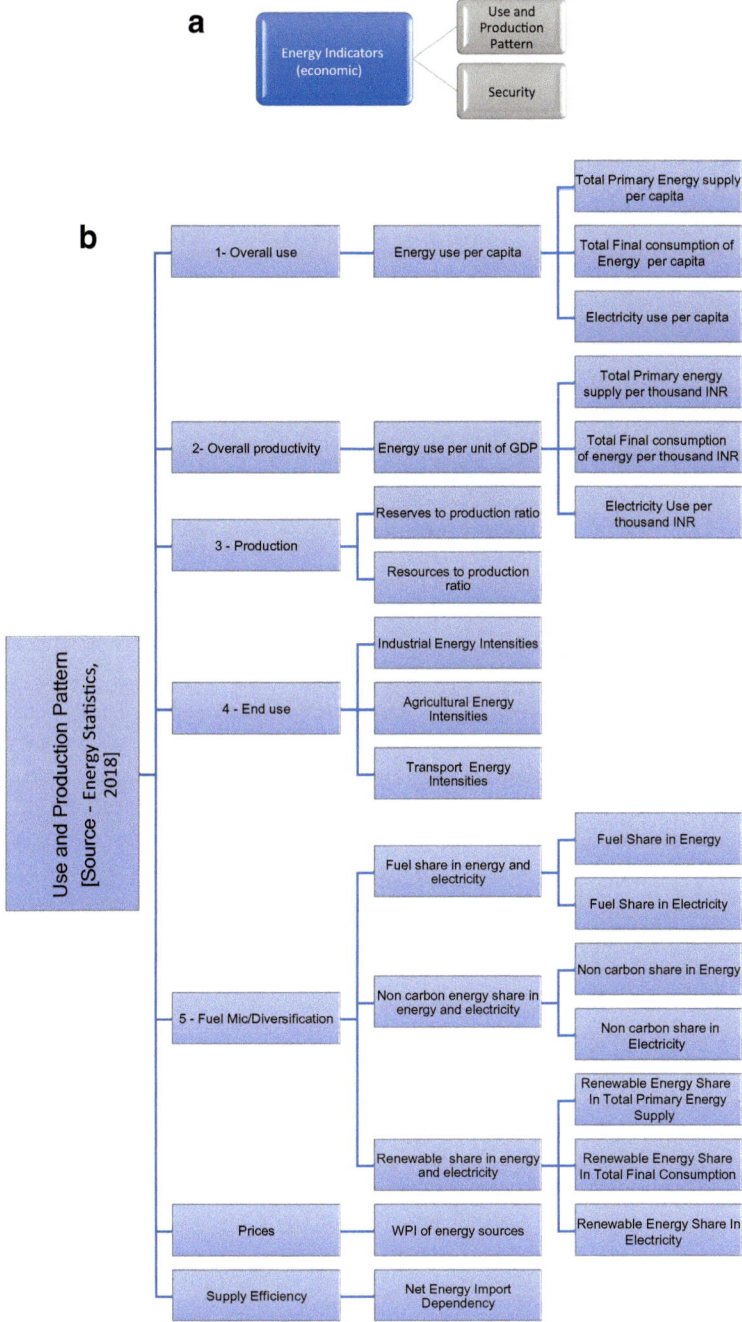

Fig. 1.3 a Energy indicators (economic). **b** Use and production pattern (*Source* Energy Statistics 2018)

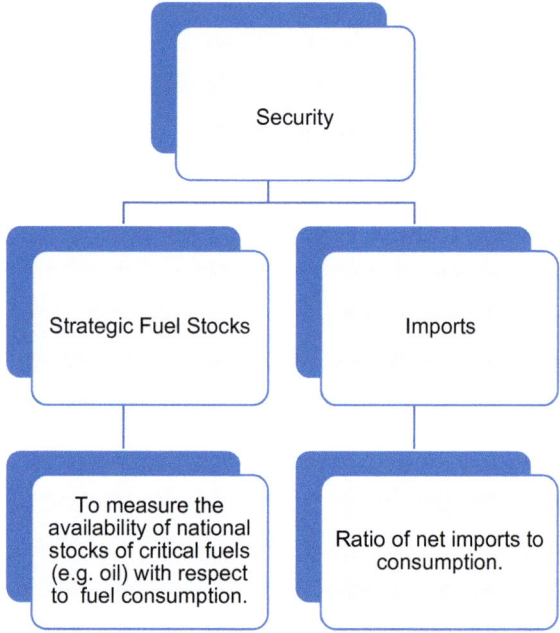

in power stations which is then transmitted through wire connected with households and commercial establishments. These power stations, in turn, use different products of nature such as water, coal, nuclear, diesel, and so on for generating energy. The third source of external energy comes from batteries such as lead acid, nickel metal, and lithium ion that are used in vehicles, computers, mobile phones, and so on.

The source of external energy comes from matter such as water, air, sunlight, petroleum, coal, biomaterial, radioactive material, and so on. Chapters ahead in the book will deal with various aspects of external energy. Both internal (natural) and external (converted or transformed) energies complement each other. For example, an underfed, unhealthy, and less educated population will fail in improving the GDP even though they have adequate supply of external energy and superior technology. In other words, the sustainable development of a nation will depend on quantitative and qualitative growth and internal as well as external sources of energy. Sustainable development in any and every nation's socioeconomic parameters is possible only when it is based on secure energy infrastructure. It is only a matter of time that the external energy infrastructure would become infructuous in the absence of strong internal energy foundation that can only be ensured through protection, conservation, and preservation of natural forests, farm lands, aquatic resources, and air which are so vital to protect and enrich our soil, ensure the availability of water and food, and stabilize our atmosphere. The concept of GDP in economic development has long been seen as a crude benchmark for measuring the value of goods and services. That is why many nations are now advocating prioritizing quality of life over financial growth (that has destroyed more than it has created). The opponents of GDP-linked economic growth argue that the current standard means of measuring economic growth ignores other factors vital to the wellbeing of the individual and populations such as environmental protection, and sustainable happiness of inhabitants. They argue that besides economic development, there is a need for measuring levels of satisfaction and community interrelationships.

Is energy (commercial) a necessity or basic need? This question has been broadly dealt in the 1987 report 'Our Common Future' prepared by the World Commission on Environment and Development (Brundtland et al. 1987). The report defines sustainable development as

> *development that meets the need of present without compromising the ability of future generations to meet their own needs*

Simply explained—sustainable development is a process of change in which exploitation of resources, technological development, financial investment, and institutional change are in harmony and enhance both current and future potentials to meet the human needs and aspirations.

The relationship between sustainable development and energy security has two important features. One is adequate energy supply that acts as a source of prosperity that satisfies the basic human needs, improves social welfare, and enhances economic growth. The other is that the production and use of energy should not endanger the quality of life of current and future generations and should not exceed the carrying capacity of ecosystems.

Sustainable development with secure energy is possible with the following options:

1. Assured availability of energy to the consumer at all times in adequate quantities, in various forms, and at affordable prices over a long-term period.
2. Increased reliance on renewable energy sources.
3. More efficient use of energy especially at end user level—in heating and cooling buildings, electrical appliances, vehicles, and industrial production systems.
4. Development and deployment of advanced energy technologies (for fossil fuels) at an accelerated pace, especially those that help in near zero harmful emissions.

Unfortunately, more than three decades after the publication of this report, developing world including India continues to rely heavily on non-renewable sources (coal and crude oil in particular), and introduction of advanced energy technologies is much slower than expected.

Energy needs for growing population and economic development are central to the planning process, particularly in oil importing developing countries like India. Uncertainty about fossil fuel prices, the growing risk of fossil fuel dependence, the high levels of GHG emissions, and the cost of alternative energy sources will ensure that wood for fuel remains in great demand at least for the next 20–30 years. Unlike other sources of domestic energy wood is easy to grow, harvest, and use with minimum technical and financial inputs as well as least reliance on outside agencies. Globally, the number of people dependent on biomass resources as their primary fuel are likely to increase to 2.7 billion in 2030 (Broadhead and Killmann 2008) with substantial increases in Asia and Africa. Since wood fuel will continue to be the main source of fuel in poor developing countries, the excess demand will be necessarily met from the trees standing on non-forest wood lands in order to buffer the impact on natural forests and defer the process of extinction.

A 'business as usual' scenario of rapid economic growth and industrialization may result in further environmental damage, and that most of the developing and underdeveloped regions may become more degraded, less forested, more polluted, and less ecologically diverse in the future (Srivastav 2019). Natural forests, in their present state of composition, productivity and extent, are in no position to meet the growing wood fuel demand and in the event of continuous extraction of wood for fuel, these forests will lose their naturalness and integrity.

Human development report of United Nations Development Program (UNDP) for the year 2016 (UNDP, HDR 2016) severely criticizes the pattern of global development that has taken place over the past three decades. The report mentions that while progress on human development front has been impressive and rapid strides have been made during the past three decades, but unfortunately the development has not been universally distributed across nations, societies, and ethnic group. There exist deep fissures in rural and urban areas, men and women, poor and rich. One of the focal areas of human development report is the HDI or Human Development Index which is based on three fundamental issues of human development, namely:

i.   A long and healthy life which is measured by life expectancy at birth.
ii.   Access to knowledge which is measured by average number of years of education received in a lifetime by people aged 25 years and older and access to learning and knowledge by expected years of schooling for children of school-entry age; and
iii. A decent standard of living measured by Gross National Income (GNI) per capita expressed in constant 2011 international dollars converted using purchasing power parity (PPP) conversion rates.

A new concept (*using both internal and external energy*) was introduced by UNDP during 2010 that was based on the deprivations suffered by households in education, health, and living standard (Fig. 1.5). Ten indicators were used to measure these three deprivations, and a cutoff at 33.3% was fixed in each case to distinguish between poor and non-poor. Of the ten indicators, education and health had two indicators each, whereas living standard had six indicators. In other words, if a household's deprivation score was 33.3% or higher, it was categorized as multidimensionally poor. Households with a deprivation score between 20 and 33.3% were categorized as living near multidimensional poverty. For the purpose of determining poverty, the report used India's data for the year 2005–2006 and concluded that 55.3% of India's population (642,391,000 people) was multidimensionally poor and another 18.2% (212,018,000 people) lived near multidimensional poverty.

A closer examination of the above indicators reveals that if a country has to improve its HDI ranking and index, it will necessarily have to ensure adequate energy supply to its citizens. The term energy in the context encompasses energy for internal growth of individual, that is derived from food, water, and air as well as for

(Source-UNDP, HDR 2016]		
AREA	INDICATOR	PERCENT SCORE
Education	School attainment: no household member has completed at least six years of schooling.	16.7
	School attendance: a school-age child (up to grade 8) is not attending school.	16.7
Health	Nutrition: based on body mass index for adults and by the height-for-age score	16.7
	Child mortality: based on death of a child five years prior to the survey.	16.7
Living Standard	Electricity: No access to electricity.	5.6
	Drinking water: No access to clean drinking water or having access through a source that is located 30 minutes away or more by walking.	5.6
	Sanitation: No access to improved sanitation facilities or having access only to shared improved sanitation facilities.	5.6
	Cooking fuel: using dung, wood or charcoal.	5.6
	Having a home with dirt, sand or dung floor.	5.6
	Assets: No access to at least one asset of information (radio, television or telephone) or having at least one asset related to information but not having at least one asset related to mobility (bike, motorbike, car, truck, animal cart or motorboat) or at least one asset related to livelihood (refrigerator, arable land or livestock).	5.6

Fig. 1.5 *Source* UNDP, HDR (2016)

external purpose such as for cooking, transportation, comfort, production of goods and services, and so on.

In this report India's HDI value is 0.624 in the order of merit and it has been positioned at 131 (out of 188 countries) during the year 2015. While India's substantial progress between 1990 and 2015 has improved the HDI value from 0.428 to 0.624, respectively, this is by no means an achievement one should be proud of. At the same time, India faces serious environmental challenges as well. India ranks 120 (among 122 countries) on the global index of water quality and 12 of the world's 15 most polluted cities are in India.[1]

For ensuring secure energy to its people, a nation must consider three important aspects, viz., adequate supply of energy (supply adequacy), availability of energy at affordable price (price affordability), and sustained supply (reliability of energy source and supply). The greatest challenge for any nation is to ensure

[1] Reference Special Report, The Economist, October 24, 2019.

all three components in the long term, especially for those countries that are net importers of energy. Such countries are always at the risk of international pressures/fluctuations/disruptions, such as price, war, political dynamics, and natural disasters. In other words, a net energy importing country cannot consider itself safe or energy secured. As far as India is concerned, it falls in the category of net importer and there is no way India can turn the table at least till the turn of this century.

Energy security in the current global context means assured supply of energy to the consumer under any circumstances at affordable prices. The concept of energy security or secure energy supplies dates to 1970s when the OPEC imposed embargoes on oil supplies resulting in supply disruptions and price fluctuations. Many developing and underdeveloped countries were taken by surprise. Before 1970, coal, natural gas, and oil were the major sources of energy in most of the countries. While proportion of coal remained almost the same during 1970s and 1980s (i.e. 32.3 and 30%, respectively), the proportion of natural gas went up from 18 to 20% during the same period and petroleum went down from 47.6 to 45.8%. International oil politics by OPEC for seven years between 1972 and 1979 sharply increased crude oil prices. From less than 3 USD in 1972, oil went up to USD 13 per barrel in 1978 and by July 1979 it was more than USD 20 per barrel (World Development Report 1979).

After the end of Second World War and after the formation of OPEC (Organization of Petroleum Exporting Countries), the world was divided into several groups (World Development Report 1980; World Bank 1980):

1. *Industrialized countries*: This group consisted of countries that were members of the Organization for Economic Cooperation and Development (OECD)[2] but excluding Greece, Portugal, Spain, and Turkey, which were included in the list of middle-income developing countries.
2. *Developing countries*: This group had two sub-groups. The low-income countries whose GNP per person was USD 360 or below and the middle-income countries whose GNP per person was USD 360 and above.
3. *Oil exporting developing countries*: This group comprised Algeria, Angola, Bahrain, Bolivia, Brunei, Congo, Ecuador, Egypt, Gabon, Indonesia, Malaysia, Mexico, Nigeria, Oman, Syria, Trinidad and Tobago, Tunisia, Venezuela, and Zaire.
4. *Capital surplus oil exporting countries*: This group comprised countries like Iran, Iraq, Kuwait, Libya, Qatar, Saudi Arabia, and the United Arab Emirates.
5. *Oil importing developing countries*: This group comprised developing countries not classified as oil-exporting developing countries or capital surplus oil exporters.
6. *Centrally planned economies*: This group included the communist bloc nations of Albania, Bulgaria, China, Cuba, Czechoslovakia, the German Democratic Republic, Hungary, the Democratic Republic of Korea, Mongolia, Poland, Romania, and the USSR.

---

[2]Organization for Economic Cooperation and Development (OECD) members are Australia, Austria, Belgium, Canada, Denmark, Finland, France, the Federal Republic of Germany, Greece, Iceland, Ireland, Italy, Japan, Luxembourg, the Netherlands, New Zealand, Norway, Portugal, Spain, Sweden, Switzerland, Turkey, the United Kingdom, and the United States.

7. The Organization of Petroleum Exporting Countries (OPEC) comprise Algeria, Ecuador, Gabon, Indonesia, Iran, Iraq, Kuwait, Libya, Nigeria, Qatar, Saudi Arabia, the United Arab Emirates, and Venezuela.

Securing resources (mainly coal, gas, and oil) for generation of electricity and for transportation during energy stress periods was the top priority for most of the energy-deficit nations including India (Figs. 1.6 and 1.7). In order to keep the cost within affordable limits, most of the countries started exploiting their own resources first. For example, the oil embargo by OPEC in 1973 and Iranian revolution in 1979 shot up coal production in developing countries (Fig. 1.8) which increased by 7.6% annually between 1973 and 1976 as compared with 3% per year in industrialized countries. More than 90% of the increased coal production was from countries with large and established coal industries, such as India, the Republic of Korea, Turkey, Yugoslavia (now Serbia and Croatia), and Vietnam. Similarly, a few countries from the underdeveloped and developing world including Argentina, China, and India started using nuclear power for energy generation while others such as Pakistan, Brazil, Iran, and Mexico, joined the club subsequently. Besides, coal, oil, and nuclear energy, countries with enough financial and natural resources started constructing hydroelectric plants. Since the gestation period of hydro plants is 5–10 years, the focus continued to be on coal and oil supply.

The decades of 1970s and 1980s depended on the following factors for energy security:

1. **Availability** of resources within the country or from outside, in turn, depended on fluctuations in the demand and supply, economic growth, changes in prices

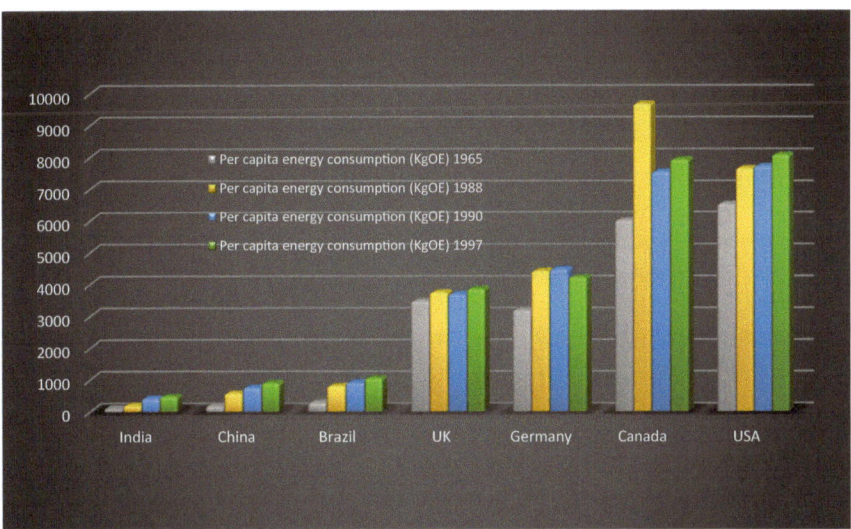

Fig. 1.6 Three decades of energy growth. *Source* WDR (1979, World Bank 1979); WDR (1990, World Bank 1990) and WDR (2000, World Bank 2000)

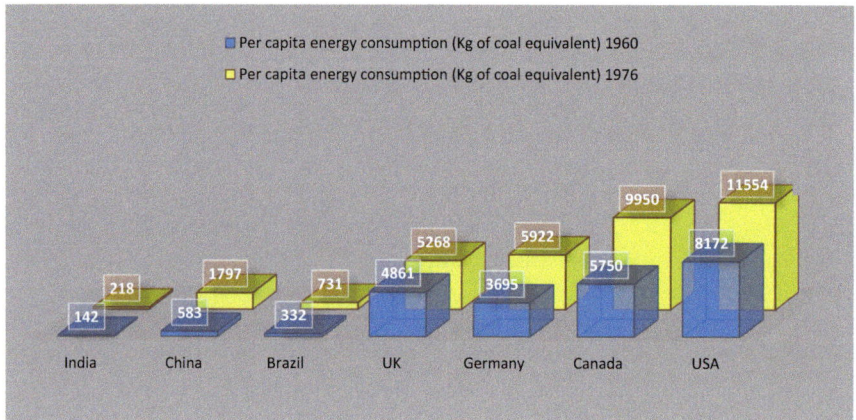

Fig. 1.7 Growth in per capita energy consumption (kilogram of coal equivalent) between 1960 and 1976 (World Development Report 1979, World Bank 1979)

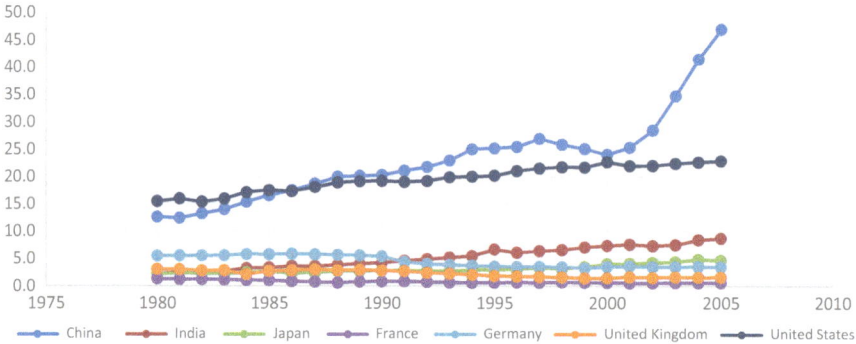

Fig. 1.8 Global coal consumption between 1980 and 2005 (in quadrillion BTU) (Ref. Brown 2008)

and incomes, as well as political and environmental factors. From 1859 onward, when the first modern well was drilled in Pennsylvania, oil changed the trajectory of global economics and geopolitics. Oil-rich countries and companies dictated market dynamics by controlling the supply of oil. Production by OPEC was the critical balancing factor as changes in their output significantly altered the global energy balance.

2. **Ease of access**, that is, the ease and convenience of accessing the resources. For resources to be procured from other countries, it was necessary to have good international as well as bilateral relations.

3. **Affordability** of resources imported varied from country to country based on resource quality as well as willingness to sell.

4. **Acceptability** to whether use of resource was acceptable socially and environmentally. While there was concern for use for coal in developed countries, but

for countries that had high import dependence, this criterion had little relevance. Acceptability of nuclear energy was another serious concern.

5. **Substitution**—possibilities of substitution of oil, gas, and coal to hydro, nuclear, and solar energy. The fact that substitution requires large investments with long gestation period which developing and underdeveloped countries could ill afford. For example, the hydropower development was deferred in many countries due to high capital cost, environmental issues, riparian state conflicts, and so on. Silent valley project in India is a glaring example of cancellation of hydroelectric project in favor of biodiversity conservation.

6. **Uncertainty** on the future economic growth of the industrialized countries was the most crucial factor for certainty/uncertainties on the demand side. Most of the OECD countries had high energy consumption (Figs. 1.9 and 1.10), and small changes in their rate of growth substantially impacted global demand for energy (Energy Information Administration 2004).

7. **Energy conservation policies** in certain key oil-importing nations were the key to energy demand and pricing. There were the possibilities for increased use of oil substitutes, particularly coal and nuclear power, and, in the long term, there was potential for using resources such as shale oils, tar sands, wind, and solar energy.

During the decades of 1980 and 1990, the global economy grew at a fast pace and it was anticipated that higher demand for energy sources could outstrip supply resulting in higher price and constraints for developing and underdeveloped world. In order to ensure secure and sustained supplies for future with minimum setbacks, countries started looking for oil substitutes such as shale oil, tar sand, and solar energy, as well

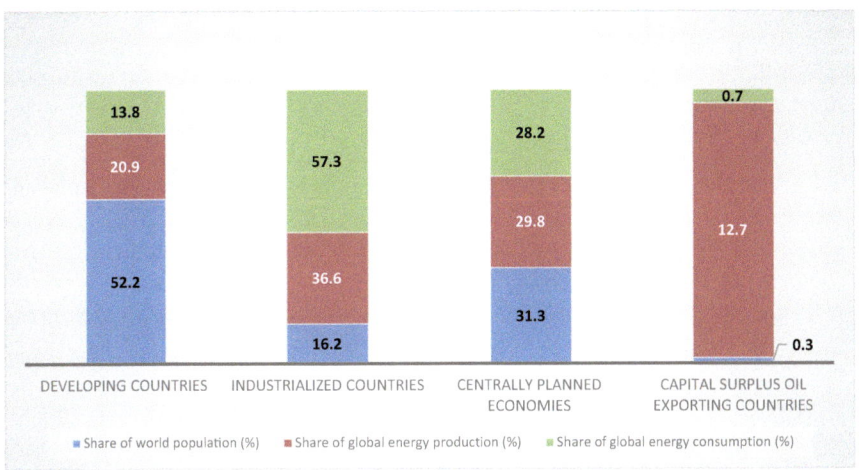

Fig. 1.9 Global energy scenario in 1976 (World Development Report 1979, World Bank 1979; World Development Report 1980, World Bank 1980)

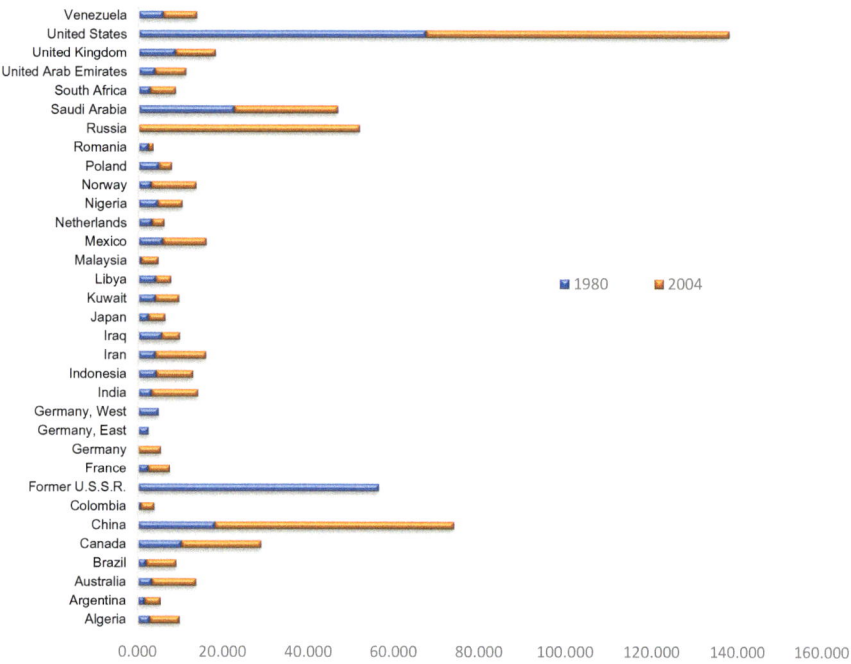

Fig. 1.10 A comparison of primary energy production (in Quadrillion Btu), in a few countries between 1980 and 2004 (*Source* Energy Information Administration 2004)

as energy conservation measures to slow down resource consumption. Investment in research and exploration was given priority for discovery of oil and gas reserves.

Nonetheless, the issue of energy security was a transition phase in which countries had to adjust to higher energy prices while ensuring that additional requirements were met from sources other than oil. Most of the countries faced challenges of varying order during this period. For instance,

- OPEC countries were besieged with the issue of sustained demand for exploiting their resources and price determination.
- Industrialized countries faced the dilemma of energy conservation, alternative sources, and safety of nuclear power plants.
- Oil-importing developing nations were concerned about domestic exploration and cheap renewable sources such as fuelwood and biogas.

At the same time, developing and underdeveloped nations risked huge capital for oil exploration with inadequate inhouse data and skilled personnel, limited petroleum reserves, and high investment in technology and infrastructure. Many countries in Asia, Latin America, and Africa invested in hydroelectric power plants as well but subsequently held back further development due to serious environmental, capital cost, rehabilitation, and other issues. Eventually, coal emerged as the most viable option for cheap energy production in non-oil commercial sector despite environmental and transportation concerns. Besides commercial energy, the non-commercial, traditional, and non-conventional sector played immensely valuable role in saving countries from energy catastrophe. Since production and consumption of energy from traditional sources (firewood, charcoal, plant and animal residues, and human and animal energy) was outside the purview of organized sector, it was considered as non-commercial though regularly traded or bartered. The World Development Report 1979 acknowledges that half of the global population relied on non-commercial energy for cooking food (World Development Report 1979, World Bank 1979). Many low-income countries in Asia and Africa continued to rely heavily on fuel from wood and animal waste till the turn of twenty-first century resulting in deforestation, degradation, and fuelwood shortages. It is surprising that the economists, intentionally or otherwise, have so far avoided inclusion of traditional sources of energy as commercial entity despite the fact that millions of households continue to use wood fuel, charcoal, and animal dung for cooking food and for other household activities. We must acknowledge that it is unrealistic to presume that every household will get electricity connection and sufficient supply of commercial energy in future and therefore, non-conventional energy technologies such as improved cookstoves, biogas plants, improved charcoal kilns, and so on must be brought under the purview of commercial energy.

The experience gained in energy access and security during the last four decades of twentieth century primarily due to international geopolitical and security situation as well as sensitivity to environmental issues helped the experts in overcoming many concerns. Starting from Iran–Iraq war, Iraqi invasion of Kuwait, the Gulf war, global war against terrorism, oil spills, Fukushima disaster, GHG emission, and so on, acceptance of renewable energy as a viable and better alternative has changed the security dimension. Governments and private sectors have become more receptive to investing in new resources, undertaking research and innovations and are willing to take risk. However, a roadmap to new strategies and approaches to governance, creating a need for increased collaboration between producer, policy maker, and national and international entities is required to be adopted.

Current global scenario for access to affordable and assured energy supplies will revolve around the following issues:

1. **International security**: Geopolitical experiences (Fig. 1.11a, b) indicate that wars, terrorism, internal conflicts, cyberattacks, embargoes, and so on adversely impact the production and transportation of resources. Therefore, stability of supplier nation/s is of utmost significance. In addition, stability of the region is crucial as well since almost two-thirds of oil supply are done through maritime routes, especially Strait of Hormuz and Strait of Malacca.[3] Any disruption in Middle East or maritime routes due to conflicts and embargo will severely impact both supplier and recipient. Not only oil but the supply of gas through pipelines is highly vulnerable. Russia supplies more than 25% requirement of gas to Europe via pipeline, most of it passing through Ukraine. In 2009, discord between Ukraine and Russia shut down gas supplies to Europe for two weeks (The Economist Weekly, April 5, 2014). Similarly, the threats of cyberattacks to the grid systems in the event of terrorist attack or full-scale war causing massive disruptions cannot be ruled out. Besides, advancement of digital transformation poses a challenge to the stakeholders and governments. In 2003, for example, a US blackout caused by a software bug cut energy supply to 50 million people. In 2010, a virus damaged centrifuges at Iran's Natanz nuclear fuel enrichment plant. A virus attack on Saudi Aramco's network in 2012 disabled 30,000 computers, erased hard drives, and replaced critical company data. In 2014, a nuclear plant operator's computer system was hacked in South Korea (Reference-Energy Security 2016).

2. **Diversification and self-sufficiency**: The country should have a basket of resources/supplier groups to choose from so that there is no over dependence on one resource. In the event of single or limited alternatives, one should at least have diversity of suppliers to import energy from.

3. **Infrastructure resilience and technology upgradation**: Coal, oil, and nuclear resources have been in use for more than a century and may remain in use for another 50 years if not more. Infrastructures built in the past need constant maintenance and upgradation for primarily two purposes. One is to avoid accidents and shutdown, and second is efficiency. Producer nation must ensure that the infrastructure is resilient and mining/extraction technology is upgraded from time to time. Users must ensure that newer technology for power generation, pollution control, and waste disposal is adopted as soon as possible to avoid criticism and supply disruption.

4. **Price volatility**: Resources especially oil have seen sharp fluctuations in price. There have been three major oil price crashes starting 1985–1986 when OPEC decided to increase supply. In 2008–2009 the oil prices went down suddenly after global financial catastrophe, went up again till 2015, and then fell sharply between June 2014 (USD 115 per barrel) and November 2015 (USD 45 per

---

[3]A third route, the Artic, is now opening up as ice has melted faster than predicted due to global warming impacts.

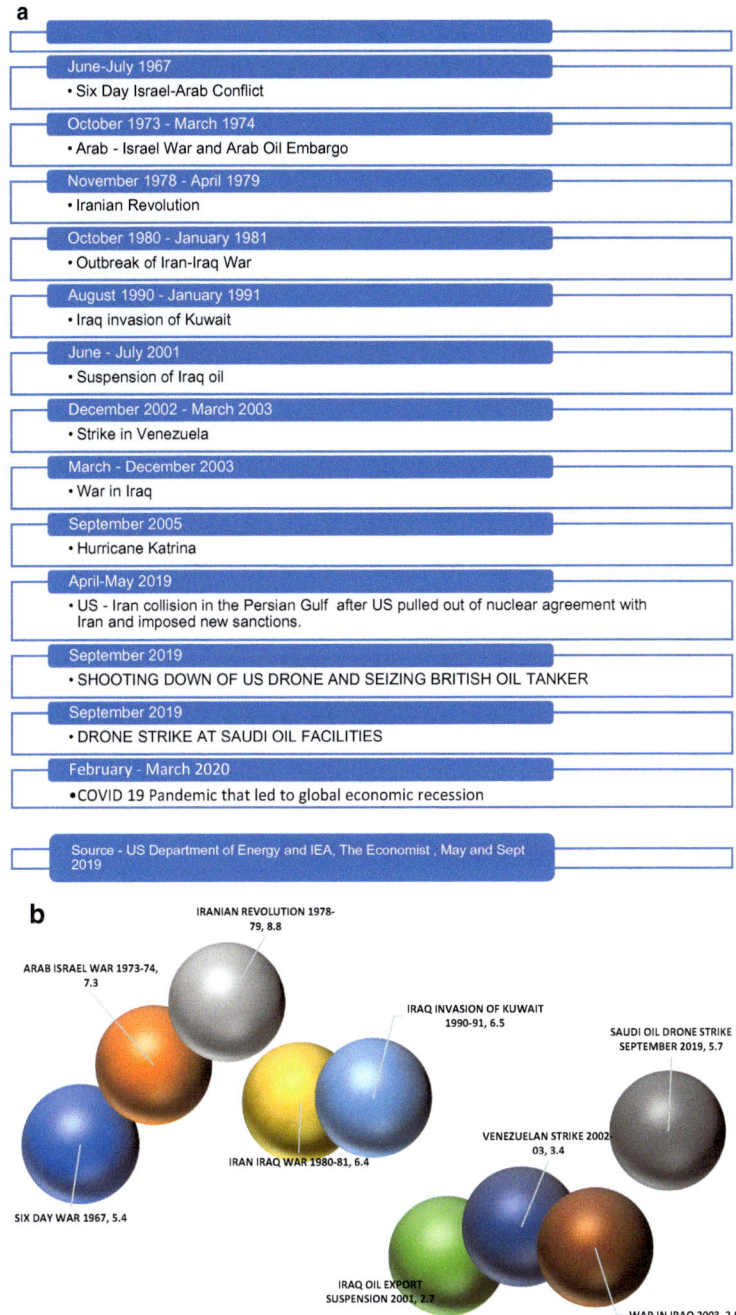

Fig. 1.11 a Noteworthy instances of oil supply disruptions. **b** Daily supply loss (Million Barrels/Day) (Ref. The Economist September 21, 2019)

barrel) and then went up. Such fluctuations affect the foreign cash reserves in fast-growing economies on the one hand and domestic production on the other.

5. **Sustainability and quality**: Since climate change impacts are real and are been felt across the globe, quality of energy especially clean, renewable, uninterrupted, and sustainable source is the only option for future. At present, there are many questions regarding sustainability of energy derived from solar and wind and other renewable sources. The technology for conversion of renewable sources to energy is evolving and expensive vis-à-vis non-renewables. Global annual investment in renewables is mere USD 300 billion annually as compared to trillions of dollars of investment in coal, gas, and oil. With almost 85% vehicles predicted to run on internal combustion engine and China and India accelerating their reliance on coal, the sustainability aspect will remain a big question till the middle of this century.

6. **Response time to transition**: Pollution, radiation effects, and climate change threats have forced the world to phase out non-renewable faster than expected and replace it with renewable sources of energy. Countries like Germany (Box 6) have decided to shut down all nuclear power plants by 2022 and is working overtime to produce solar and wind energy even though it has largest deposits of lignite in the world. Other countries including Austria, Finland, Denmark, Portugal, and France are leading the energy transition to alternative and renewable energy sources as well. Transition is not easy as it involves money and technology and therefore, societal consent and complimentary policies are significant. However, a successful transition will require market transformation and significant changes to the electric utility business model and regulatory policies.

Box 6

ENERGIEWENDE

Energiewende is a German word that basically means energy revolution. It is Germanyís most ambitious and revolutionary program of switching over from fossil and nuclear energy to renewable energy. Germany has decided to switch off its last nuclear power plant by 2022 and to increase the share of renewable energy (solar, wind and biomass) to 80% of total electricity production and 60% of overall energy use by 2050. Germany enacted a renewable energy law in 2000 that guarantees 20 years of fixed price to solar and wind energy producers as a result of which solar panel and windmills dominate Germanyís landscape. Germany has been gradually reducing use of nuclear energy but not of coal and lignite and that is a big concern for the advocates of energiewende.

(Ref - The Economist, January 18th, 2014)

7. **Waste disposal**: This is an important threat, especially safe disposal of nuclear waste. Most consumers are of the view that solar energy is the safest and best option for future not realizing that even the waste generated from solar panels contains toxic substances such as cadmium and chromium.

8. **Political relationship and arm-twisting**: More than stability of the importing nation, it is the political rapport between the countries and the region that is of paramount importance. Russia is the largest supplier of natural gas to most of the Europe (largely eastern), including countries like Lithuania, Estonia, Finland, and Latvia where it supplies 100% of their requirement as well as to Germany, Italy, France, Netherland, Belgium, and so on in lesser proportion. Most of the gas is supplied through pipeline that pass-through Ukraine. The ongoing political discord between Russia and Ukraine may endanger supplies to Ukraine by Russia or Ukraine can disrupt supplies to Europe. In fact, Russia shut down supplies to Ukraine in January 2009 for two weeks. Arm-twisting by Russia makes Europe highly vulnerable to supplies, they have increased their storage capacity, and interconnected gas pipelines between them, which in tun means additional expenditure and more risk.

9. **Emergency response**: The history of emergency whether at national, regional, or global level reveals that such exigencies occur frequently and unpredictably. Fukushima nuclear disaster is a case in point where Japanese were found wanting after the tsunami and radioactive leaks, shut down all nuclear plants, and vowed to switch over to renewables. Likewise, the US–China trade war, global economic slowdown, and COVID-19 attack have brought global economy on its knees. Badly hurt economy, mass deaths, disruption in supply of labor, goods and services, rising unemployment, and drain of savings will affect everyone in the country whether rich of poor with varying intensity. This also means deceleration in the momentum for renewable energy and setback to climate change mitigation efforts. The lesson from such events—Every nation irrespective of its economic status must prepare an emergency shield against power shut down due to natural and other disasters.

10. **Conservation and efficiency**: Continuation of fossil fuel is not only a serious energy security issue but an equally potent global warming weapon. Any policy that can alleviate one of the threats can also help to address the other. Improving energy efficiency through technological improvements can result in less consumption of fossil fuels and reduction in greenhouse-gas emissions in near future. Conservation of energy is also an effective way of mitigating climate change impacts especially where fossil fuels are used to provide electricity and for transportation. For instance, switching off the light when leaving a room or walking and cycling instead of driving. In the long run replacement of coal to gas or to other zero carbon energy sources will be must as coal emits almost 75% more carbon per unit of energy contained in the fuel than natural gas and about one-third more than oil.

11. **Energy research** in energy efficiency, nuclear fission and fusion, hydrogen and fuel cells, and storage technologies.
12. **Climate change impacts** are less talked about but are serious threat to energy security, especially to thermal and hydropower plants. Thermal power plants require water for cooling and hydel plants need it for running the turbines. Short supply of water in the event of severe and continuous drought can affect power generation. Poland suffered low power generation in August 2015 due to less water availability for cooling in thermal power plants severely impacting manufacturing sector. El Nino effect on global rainfall pattern appears to be biggest security threat to the movement of ships and containers. One glaring event in 2019 was the lowering of water level in Panama Canal. Under normal rainy season (mid-April to mid-December) the water level in lake Gatun of Panama Canal remains at 26.5 meters above MSL and falls to 25.9 meters during dry season. If the water level falls below 24.4 meters (due to drought or prolonged dry spell), movement of large container ships is stopped. In 2019, the water level went down by 1.8 meters from normal level threatening the movement of containers. This is attributed to El Nino effect that is responsible for prolonged droughts in the Panama Canal Watershed area since 2014–2015.

1.4 Energy Security and Access in India

The World Economic Forum has developed a tool to assess the performance of energy systems of individual countries using three indices, viz., economic growth and development, environmental sustainability, and energy access and security. Various indicators and sub-indicators and weightage given to each sub-indicator for energy access and security are shown in Box 7.

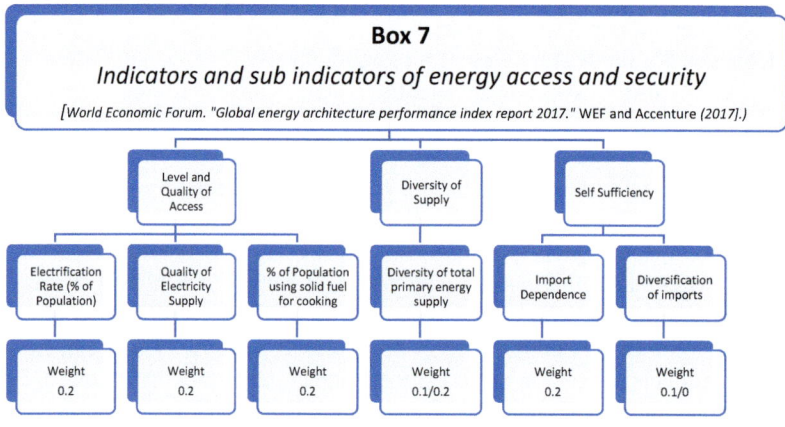

Fig. 1.12 Energy access and security score (World Economic Forum. "Global energy architecture performance index report 2017, WEF and Accenture 2017)

India is ranked 87 among 127 countries with a score of 0.62 (Fig. 1.12). India's energy system continues to face significant challenges, particularly in environmental sustainability especially carbon dioxide emissions from electricity production and pollution from vehicles, agricultural crop burning, refuse combustion, and fireworks. India also faces an uphill task of ensuring electricity access to all households and replacement of biomass-based fuels used for cooking.

Since the turn of twenty-first century, India has been poised for a big leap in its economy and therefore requires secure access to modern sources that are affordable and reliable so as to help reduce poverty, improve health, increase productivity, and promote economic growth. Besides energy security, India will have to adopt actionable strategy for decarbonization of its energy mix through:

i. *Provision of adequate amount of electricity including increased consumption over a period.*
ii. *Provision of clean cooking fuel.*

This may be possible subject to adequate investment, improved technology, and infrastructure and better governance. The conservative estimates are that close to 10% population will remain devoid of adequate electricity in India by 2030[4] (IEA, WEO 2011). A serious drawback in the existing system of providing large subsidy or free connectivity, if allowed to continue, may prove counterproductive in the long run.

As far as clean cooking fuel is concerned, India's position is precarious. In 2009, there were almost 840 (749 million rural and 87 million urban) million people in India relying on traditional biomass-based fuels. This number is expected to go down slightly to 780 million (719 rural and 59 urban) in 2030[5] (IEA, WEO 2011). There is no reason to believe that this number can be reduced to zero by 2030 or even by 2040 unless this becomes top priority for the government at state and national level and billions of rupees (average annual cost between 2010 and 2030 has been estimated to be 0.8 billion USD) are earmarked every year for biogas, off grid solar energy systems, micro-hydro systems, improved cookstoves, supply of natural gas, local capacity and operations, and so on.

India has been undergoing transformative changes since independence with population and poverty as the pivotal challenge. With the passage of every day, India faces the toughest challenge of energizing its economic development through power generation that is highly dependent on imports of raw materials. The nature of imported raw material is changing fast, from coal and oil during nineteenth and mid-twentieth century to solar and wind subsequently. As the situation stands today, the bad fuel (coal, oil, and nuclear) is expected to be replaced by cleaner fuel basket of solar wind and hydrogen fuels, at least in western Europe and North Americas.

In case of India, population will be the main limiting factor to energy security, availability, and access (Box 8). Rapid policy changes, especially privatization introduced in the aftermath of 1991 financial crisis, gave impetus to infrastructure development, rural development, telecommunication, transportation, mobile phones, internet connectivity, e-governance, and e-markets. Besides, the benefits of space technology have changed the economic landscape of India.

---

[4]Ref—[IEA, International Energy Agency. "World energy outlook 2011." *Int. Energy Agency* **666** (2011)].

[5]Ref—[IEA, International Energy Agency. "World energy outlook 2011." *Int. Energy Agency* **666** (2011)].

Box 8

What is meant by 'Access' to modern energy'?

In 2011, the IEA defined modern energy access as

'a household having reliable and affordable access to clean cooking facilities, a first connection to electricity and then an increasing level of electricity consumption overtime to reach the regional average'.

The definition of access also involved consumption of a specified minimum level of electricity that varied with the location of household (rural or urban area). The initial threshold level of electricity consumption for rural households was assumed to be 250 kilowatt-hours (kwh) per year and for urban households it was 500 kwh per year. This assumption (in consumption) for rural areas was based on the premise that the electricity would provide for the use of a floor fan, a mobile telephone and two compact fluorescent light bulbs for about five hours per day. For urban areas, the appliances included an efficient refrigerator, a second mobile telephone per household and another appliance, such as a small television or a computer. This definition of energy access also included such cooking facilities that was safe to human health, environmentally sustainable and energy efficient than the average biomass cookstove currently used in developing countries. This definition refers primarily to biogas systems, liquefied petroleum gas (LPG) stoves and advanced biomass cookstoves that have considerably lower emissions and higher efficiencies than traditional three stone cookstoves.

There was another assumption for 'accessibility' – the consumption level was expected to rise overtime to reach average consumption in the region within a span of five years.

[IEA – 2011, World Energy Outlook Special Report].

In 2017, The International Energy Agency (IEA) has redefined energy access as follows:

"A household having reliable and affordable access to both cooking facilities and to electricity, which is enough to supply a basic bundle of energy services initially, and then and increasing level of electricity overtime to reach the regional average"

The term 'basic bundle of energy services' in the above definition means several lightbulbs, a flashlight, phone charging and a radio.

The term 'access to clean cooking' in the above definition means access to modern fuel and technologies including natural gas, LPG, electricity, and biogas and improved cookstoves.

[Energy Access Outlook 2017, World Energy Outlook (Special Report)]

With the current population of nearly 1300 million, sustaining economic development will require higher rate of growth in energy demand and secure supplies. India's coal, oil, and gas imports are expected to be much higher with the passage of time. Consequently, the risk of supply disruption will also be higher mainly due to wars, embargoes, and price fluctuations. India's GDP per person remained abysmally low, below USD 450 till the end of twentieth century. At the turn of twenty-first century, India was a low-income country with per capita GDP of USD 462 in comparison with USD 911 for China, USD 1,270 for the developing countries, USD 22,149 for OECD countries, and USD 35,277 for the USA. Despite the recent rise in incomes, the average per-capita GDP of India in 2015 was 1606 USD (http://www.worldb ank.org). There is stark difference in per capita income across the country. For example, a person in Bihar, the poorest state, earns about a tenth of that in Goa, the richest state. Under the circumstances it will be irrational to expect magical rise in per capita GDP of India.

At the current rate of population growth, India is projected to be the most populous country of the world by 2030 with 1.5 billion people spread over inhabitable areas of the country's 328 million square kilometers. Most of the population among the young and middle age group will move from rural areas to peri-urban and urban areas. And more and more peri-urban areas will be carved out of agricultural land of 640,000 villages. Over 70% of the population lives is linked to over 640,000 villages in the country, contributing 50% to India's GDP. 18% of population lives in 7,834 towns contributing 25% to GDP. 12% of population lives in 44 cities (Registrar General, India 2011). The iconic pyramid has expanded in its middle, creating a 'diamond' by the increasing 'climbers' entering the lower/middle classes in both urban and rural areas.

Unprecedented population growth will outstrip the demand for oil and gas (and modest reserves) to fuel transport and power generation and insufficient domestic production will ensure import dependency. India has and will continue to rely on Middle East for its oil supply, the reason being comparative cheapness, ease of transport through sea, and geographic nearness. By 2030, India is expected to import 6–7 million barrels a day mostly from Middle East countries. In the medium term, at least, most of the Middle Eastern oil shipped to India will continue to transit the Straits of Hormuz at the mouth of the Persian Gulf—the world's busiest oil-shipping lane. India's gas requirement will be between 61 and 112 billion cubic meters by 2030 depending on economic growth and renewable energy potential with almost complete import from Middle East. India imported 12% of its total coal needs of 36 Mtce in 2005 and the same is expected to grow between 244 and 282 Mtce. Coal is relatively more expensive to transport, and mining invites sharp criticism from conservationists. Future coal demand and import will be proportional to use of clean technology and switchover renewables in the overall energy basket (Fig. 1.13). Besides several constraints in power generation, the country also faces transmission and distribution losses for a variety of reasons, viz., substantial energy sold at low voltage, sparsely distributed loads over large rural areas, inadequate investment in distribution system, improper billing, and high pilferage. India's average transmission and distribution losses exceed 25% of total power generation which is almost 2.5 times the world average.

India has adopted some measures to reduce its vulnerability to oil supply disruptions. In 1998, a new exploration licensing policy aimed at encouraging investment in the upstream oil and gas sectors was adopted. The government has also decided to

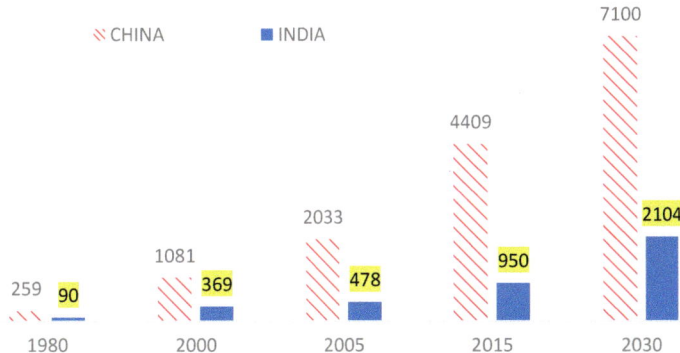

Fig. 1.13 Electricity demand (India and China) in TWh (*Source* "World Energy Outlook 2007 Edition–China and India Insights." 2007)

build a strategic petroleum reserve of emergency stocks that could play a particularly important role in enhancing short-term security. Starting with 16 days of reserves, India plans to increase it to 90 days.

The Indian government also encourages public and private companies to acquire oil assets overseas and promotes the development of natural gas, clean coal technology, nuclear power, and renewables to diversify energy use away from oil in both non-transport and transport uses. Several measures (Box 9) have been introduced to promote more efficient energy use and reduce waste, including the phasing-out of state subsidies on petroleum products higher taxes on transport fuels. Increasing emphasis is now being given to energy efficiency, and renewable energy.

Box 9 - Main policy responses in India to rising energy insecurity [IEA.WEO (2007)]	
Energy savings	1. Establishment of the Bureau of Energy Efficiency for better coordination. 2. Improved efficiency (and consequent reduction in loses) through renovation of power stations and grid. **3.** Introduction of mandatory appliance labelling and energy efficiency standards for buildings.
Diversification of resources	1. Shift to renewable resources such as wind solar and biomass in addition to existing use of water, coal, gas and nuclear in power generation. 2. Encouragement and promotion of biofuels and CNG for public vehicles in several cities. 3. Solar water heating systems and solar air heating/steam generating systems for individual /community use;
Diversification of oil supply sources/routes	1. New Exploration Licensing Policy (NELP) attracting foreign companies;
Strategic oil reserves	1. Hydrocarbon Vision 2025 encourages overseas equity investment, backed by governmentís energy diplomacy [Planning Commission, 1999]. 2. Rapid development of ONGC Videshís assets portfolio.
Equity oil overseas Acquisitions	1. Setting up Strategic Petroleum Reserve with capacity of 36 million barrels to be increased subsequently to 110 million barrels.

The International Energy Agency[6] has estimated that the global spending on energy should be in the range of 680 billion USD per year if we must improve energy security and efficiency. Simultaneously, we will have to revolutionize the energy demand of transport sector that currently consumes 27% global energy. This looks like an impossible task for developing countries where neither government have enough funds for replacing archaic power generation and distribution technology nor majority population as they are only willing to own cars that are cheap, use diesel or compressed gas, and low on maintenance. Similarly housing that consume nearly 40% of energy for lighting, heating, cooling, and ventilation will be required to be converted to 'Net Zero Energy—Zero Carbon' buildings,[7] another gigantic task to achieve (Srivastav 2019). Most of the existing buildings require better insulation, proper ventilation, and natural lighting, as well as improvements in space and water heating. Unfortunately, in the absence of energy efficiency codes in many countries, the future of 'Net Zero Energy—Zero Carbon' buildings does not appear bright at least in developing countries where simple things like judicious placement of windows and roof shadings, natural green surroundings, high ceilings and vents, efficient in house lighting, and placing solar panels on roof tops are overlooked in even new commercial and residential buildings.

References

Aayog NITI (2017) Draft national energy policy. National Institution for Transforming India, Government of India, New Delhi. http://niti.gov.in/writereaddata/files/new_initiatives/NEP-ID_276

Birol F (2007) World energy outlook 2007: China and India insights. Council on Foreign Relations, Inside CFR Events podcast, MP3 file 1: 02-17

Broadhead J, Killmann W (2008) Forests and energy: key issues, No. 154. Food & Agriculture Org

Brown LR (2008) Plan B 3.0: mobilizing to save civilization (substantially revised). WW Norton & Company

Brundtland GH, Khalid M, Agnelli S, Al-Athel S, Chidzero B (1987) Our common future. New York 8

Chen W-Y, Suzuki T, Lackner M (eds) (2017) Handbook of climate change mitigation and adaptation. Springer International Publishing

Cherp A, Jewell J, Vinichenko V, Bauer N, De Cian E (2016) Global energy security under different climate policies, GDP growth rates and fossil resource availabilities. Climatic Change 136(1): 83–94

Davis GH, Wood LA (1974) Water demands for expanding energy development, vol 703. US Geological Survey, National Center

Eckholm E (1975) The other energy crisis: firewood. Worldwatch Paper 1

Energy Information Administration (EIA) (2004) International energy annual 2004

Energy Statistics (2018) Central Statistics Office, National Statistical Organization (Ministry of Statistics and Programme Implementation, Government of India, 2018)

---

[6]Ref—EIA (2004).

[7]A 'Net Zero Energy—Zero Carbon' building is the one that produces 100% required energy through renewable on site and therefore emits no CO_2.

Foresti G, Guizzo S, Trenti S (2010) Environmental policy, technology and trade in environmental goods: what about China. Intesa Sanpaolo, Venice

IEA (2007) World energy outlook 2007 edition–China and India insights

IEA (2017) Energy access outlook 2017, World energy outlook special report

IEA, International Energy Agency (2011) World energy outlook 2011. Int. Energy Agency 666

IEA, International Energy Agency (2011) World energy outlook special report. Int. Energy Agency

Mason M, Mor A (eds) (2009) Renewable energy in the Middle East: enhancing security through regional cooperation. Springer Science & Business Media

Outlook, India Energy. World energy outlook special report 2015/International Energy Agency. https://www.iea.org/publications/freepublications/publication/IndiaEnergyOutlook_WEO2015.pdf

Parr A (2011) Hydraulics and pneumatics: a technician's and engineer's guide. Elsevier

Planning Commission (1953) First five year plan

Planning Commission (1999) Hydrocarbon vision 2025

Registrar General, India (2011) Census of India 2011: provisional population totals-India data sheet. Office of the Registrar General Census Commissioner, India. Indian Census Bureau

Srivastav A, Srivastav S (2015) Ecological meltdown: impact of unchecked human growth on the earth s natural systems. The Energy and Resources Institute (TERI)

Srivastav A, Srivastav S, Nishida (2019) The science and impact of climate change. Springer

Uhlmann B (2015) EU law and policy on energy security. Streamlining of environmental impact assessments for cross-border electricity infrastructure. PhD dissertation

UNDP, HDR (2016) Gender inequality index. Human development report

UNDP, UNDESA (2000) WEC (2000) World Energy Assessment: energy and the challenge of sustainability. United Nations Development Programme, New York

Wesley M (ed) (2007) Energy security in Asia, vol 3. Routledge

World Development Report, 1979. World Bank (1979)

World Development Report, 1980. World Bank (1980)

World Development Report, 1990. World Bank (1990)

World Development Report, 2000. World Bank (2000)

World Economic Forum (2017) Global energy architecture performance index report 2017. WEF and Accenture

Yearbook, Energy Statistics (2013) New York: United Nations, Department of Economic and Social Affairs. Statistics Division

Zou C (2020) New energy

http://www.worldbank.org

https://www.quora.com/What-is-the-difference-between-metric-ton-short-ton-and-long-ton

Chapter 2
Energy Sector Progression in India

Abstract India's economic and industrial expansion has not been commensurate with rapid population explosion after independence. With abysmally poor per capita consumption and limited indigenous sources for energy generation, India committed huge quantum of its financial and other resources for food, water, and energy security. Over a period of seven decades that witnessed numerous crests and troughs of energy insecurity especially from global forces coupled with national and international environmental concerns for power projects, India has been able to connect almost all rural and urban areas in its electricity network. All five-year plans and interregnums planned and executed by the Government of India gave priority to power sector in setting up hydro, thermal, and nuclear plants in a phased manner. Nonetheless, the imminent global warming threats have forced every nation to relinquish non-renewables and look for other options including solar, wind, and geothermal. This chapter is a synopsis of India's journey in energy sector post-independence.

Keywords Coal · Natural gas · Petroleum · Nuclear · Planning commission

Largest among all South Asian nations in terms of population, land area, and coastline, the triangular-shaped Indian peninsula covers an area of 3.28 million km². Geographically, India can be divided into five parts, namely:

i. The Himalayan range in the north;
ii. The Northern Plains;
iii. The Great Peninsular Plateau;
iv. The Coastal Plains; and
v. The Islands.

There are 14 major, 44 medium, and 55 minor rivers that cumulatively discharge about 1566 thousand million cubic meters of water annually through land drainage into the seas. Nine of 14 major rivers meet the sea in the east coast and the remaining five in the west coast. The coastline of the eastern (Bay of Bengal) coast differs from the western (Arabian Sea) coast. The east coast is characterized by a narrow continental shelf and slow-moving rivers with four major deltas formed by the Cauvery, Krishna, Godavari, and Mahanadi. India has often been described as a rich land

© Springer Nature Singapore Pte Ltd. 2021   33
A. Srivastav, *Energy Dynamics and Climate Mitigation*,
Advances in Geographical and Environmental Sciences,
https://doi.org/10.1007/978-981-15-8940-9_2

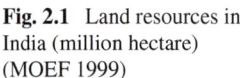

Fig. 2.1 Land resources in India (million hectare) (MOEF 1999)

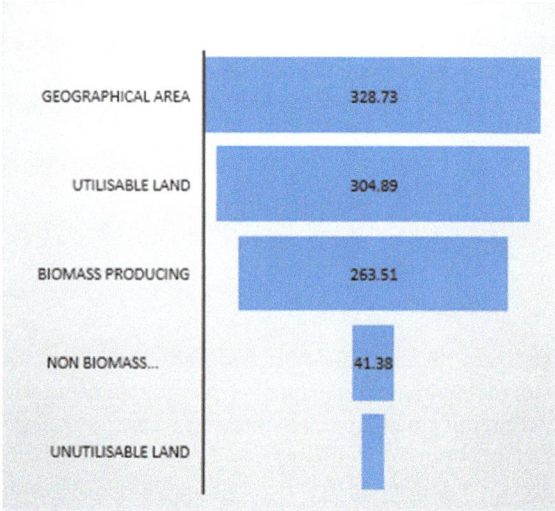

with poor people. Of the 329 million hectares landmass, some 263 million hectares (Fig. 2.1) are potentially available for human use of which the agricultural production mainly comes from 170 million hectares (MOEF 1999). The remainder includes natural forests and protected areas covering about 74 million hectares, which are responsible for all provisioning services. Perpetual sunshine ensures high level of biological productivity, nutrient recycling, and soil water retention. This possibly explains how 2.4% of the world surface area can meet the growing demands of more than 16% of world's human and associated livestock populations.

After independence, India's rapid population, economic and industrial growth exerted enormous pressures on its resources contributing significantly to the degradation of the natural ecosystems. Currently, around 40.27 million hectares are prone to cyclones and floods. Such extreme conditions and frequent disasters have greatly eroded developmental gains. The coastal region is subjected to ecological stress and environmental conflicts due to erosion, overexploitation of natural resources, loss of biological cover, increased salinity, ground water depletion, and pollution. With continuously increasing population pressure in the coastal states, the vulnerability to frequent and intense natural and manmade hazards is on the rise.

Food and energy security are two major issues that have caused concern to millions of Indians over last several decades. Being a poor underdeveloped economy at the time of independence, India focused its resources on basic human needs such as food, shelter, fuel, clothing, safe water, healthcare, and education. There was hardly any concern to understand the energy–poverty–environment–health nexus till the global warming and climate change started threatening the very existence of humanity, especially in terms of providing access to adequate, affordable, reliable, high quality, safe, and environmentally benign energy services to support economic development and human welfare.

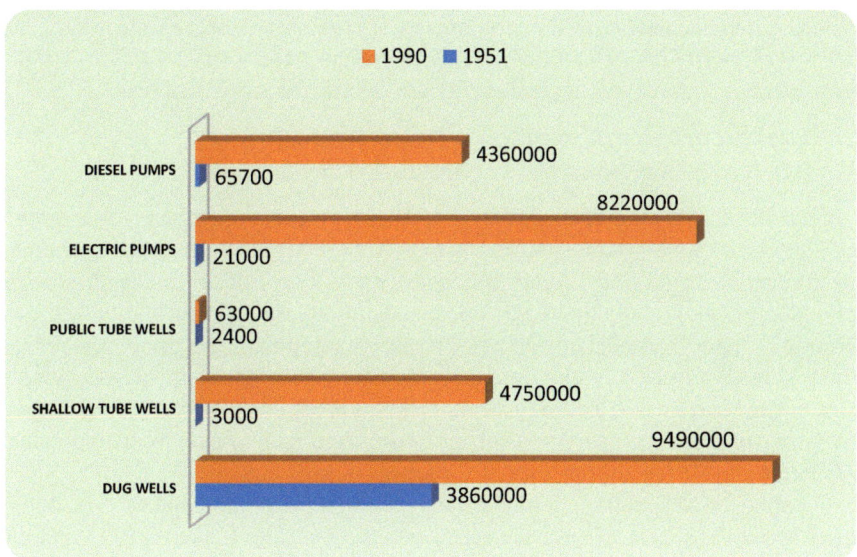

Fig. 2.2 Ground water extraction efforts. *Source* UNICEF (1998)

Green revolution during 1960s and 1970s undoubtedly improved the food grain production but at an enormous cost. It led to a massive shift in irrigation from surface water to ground water in the initial few decades (Fig. 2.2) (UNICEF 1998).

The proportion of cultivated area irrigated by ground water increased from about 30% in 1965–1966 to over 50% now. Nonetheless, the water table has gone down rapidly, and many bore wells have failed. India has paid exorbitant price for enhancing the productivity potential of its arable land. Sea water intrusion into the aquifers is now widespread and eroded soils lose around 10–300 mm/ha/year of water. Post green revolution India produces around 200 kilos per head per annum of food grain, which is far from being satisfactory. Most of the poor cannot afford to buy these and still rely on the coarse grains for health and energy needs. The situation on energy front, especially commercial, is worrisome. India has been using coal (a major source of carbon dioxide) for more than a century now and the existing data indicates that there is no substitution in sight for next several decades.

2.1 Energy as a Commercial Entity

Energy as a commercial entity was an uncommon word in India during the nineteenth century since most of the people relied on wood fuel, plant as well as animal waste for meeting their cooking, heating, and lighting requirements. Wood was available aplenty inside and outside the vast tracts of forest lands and livestock rearing was easy—almost free of cost—as fodder was plentiful and animals were free ranging.

Indians got conversant with the term energy after the enactment of electricity law that was primarily enacted for use and supply of electrical energy. Section 2 of the India Electricity Act 1910 defined energy as 'electrical energy' that was.

I. Generated, transmitted, or supplied for any purpose, or
II. Used for any purpose except the transmission of a message.

Until recently, the Indian Electricity Act (1910), the Electricity Supply Act (1948), and the Electricity Regulatory Commissions Act (1998) were the main regulations for the energy sector. Since commercial energy was a precious resource, these laws prima facie dealt with the transmission, consumption, use and control of electricity through licensing mechanism and not with the generation of electric power per se. There were severe penal actions for theft of electricity. In 2003, all previous laws were consolidated and a new Electricity Act (2003) was brought into force (Bhattacharyya 2005). Apart from the national level acts, each state is governed by its individual legislations.

For almost four decades after independence, energy sector continued to be dominated by public sector, and private sector had limited role. It was in 1991 that the policy on private participation in the power sector was adopted to encourage private sector participation.

At the time of independence (India was an underdeveloped economy then), around 68% of the population was engaged in agriculture, around 14% in industry (large and small scale), 8% in trade and transport, and the remaining 10% in professions and services. The canals in northern India discharged 400,000 cusecs[1] of water to irrigate 24 million acres of land in undivided India. Post 1947, almost half of the canal system and water supply went to Pakistan and this led to serious food scarcity in India. To add to the woes, large part of the country lacked basic services such as power, irrigation, transportation, and communication. The size of agricultural holdings had declined over the previous five decades, old cottage and small-scale industries were decaying, and the rural population that constituted about 83% of the total suffered from chronic underemployment and low incomes since farming in India was heavily dependent on south-west monsoon (even today, agriculture continues to be dependent on monsoon despite large number of multipurpose hydroelectric projects and massive canal networks for irrigation, farming, and farm labor remain largest source of jobs and account for 1/6th of India's GDP[2]). Food security was an important area as the death of millions of Indians during the famines of 1876–1877 and 1943 (Bengal famine) was permanently registered in every Indian's psyche.

After independence, India was divided into following states (Box 1).

---

[1]Cusec—A unit of flow of water equal to one cubic foot per second.
[2]Reference—The Cloud Messenger. The Economist June 29, 2019.

Box 1 - INDIAN STATES DURING THE FIRST FIVE YEAR PLAN [Source Plan. Comm 1953]			
PART A	**PART B**	**PART C**	
1. ASSAM	1. HYDERABAD	1. AJMER	1. J & K
2. BIHAR	2. MADHYA BHARAT	2. BHOPAL	
3. BOMBAY	3. MYSORE	3. BILASPUR	
4. MADHYA PRADESH	4. PEPSU	4. COORG	
5. MADRAS	5. RAJASTHAN	5. DELHI	
6. ORISSA	6. SAURASHTRA	6. HIMACHAL PRADESH	
7. PUNJAB	7. TRAVACORE-COCHIN	7. KUTCH	
8. UTTAR PRADESH		8. MANIPUR	
9. WEST BENGAL		9. TRIPURA	
		10. VINDHYA PRADESH	

The independent India adopted a coordinated approach for electricity during post-war (Second World War) reconstruction and development of the country, including provision for supply of electricity to semi-urban and rural areas. The Indian constitution incorporated electricity in the concurrent list, thereby ensuring share of responsibility for both the federal and the state governments for this sector.

There were three predicaments that India faced. One, it was not richly endowed with the then known commercial sources of energy, viz., coal, natural gas, and petroleum; two, the overall efficiency of thermal power stations was much below average (in 1959–1960, the efficiency was 19.5%) as the units were small and old; and three, the population of India increased at an unprecedented rate during the first half of twentieth century reaching 356 million in 1951 from 235 million in 1901. The growth in electricity generation could not keep pace and the same is reflected in the rural as well as urban electrification of the nation. Most of the villages below the population of 10,000 individuals did not have access to electricity in 1951 as is evident from Fig. 2.3.

India started generating its commercial energy from three sources, namely coal, petroleum, and water. Most of its coal was in Bihar and western India with small deposits in Assam, Madhya Pradesh, and Hyderabad. Besides coal, large deposits of lignite were found in South Arcot district of Madras and in Kutch. Oil production was insignificant, and India was then producing only 5% (Assam oil field) of its total requirement of petroleum and the balance 95% was imported. Fortunately, India had massive hydro resources with an estimated potential of 40 million kilowatts.

It will be worthwhile to mention that the initial phase of commercial power installation in India started in late nineteenth century in the form of coal-based thermal power stations when the first station was set up in Calcutta (now Kolkata) in 1879 followed by one in Bombay in 1882. Subsequently, the first hydroelectric station in

Population range (1941 Census)	Total number of villages /towns	Number of villages /towns with public electricity supply	Percentage of towns or villages with public electricity supply to total
Over 100,000	49	49	100.00
50,000 -- 100,000	88	88	100.00
20,000 --- 50,000	277	240	86.64
10,000 -- 20,000	607	260	42.83
5,000 -- 10,000	2367	258	10.86
Below 5,000	559062	2792	0.50

Fig. 2.3 Access to electricity in 1951 (*Source* Planning Commission 1953)

India was set up at Sivasamudram in Mysore in 1902 followed by Tata hydroelectric station in Bombay (now Mumbai). Other large cities were brought under the ambit of electrification in next two decades. As time progressed the electricity generation got improved and by 1939, the generation capacity had reached one million kilowatts. Of this nearly 60% was generated by thermal power stations, 8% by oil burning stations, and 32% by hydroelectric stations. The net electricity-generating capacity in the country in 1950 was approximately 2.3 million kilowatts, of which 1.7 million kilowatts was from thermal stations (including oil burning plants) and about 560,000 kilowatts from hydroelectric plants. Most of the thermal power stations were set up close to coal mines in Bihar and West Bengal and the states of Punjab, UP, Bombay, and South India were supplied power through hydropower stations.

Commercial energy played a significant role in improving farm production by supplying electric power for large-scale installation of tube wells and lift irrigation system. Nearly 35% of the electric energy generated (about 61 million kilowatt-hour) by the Ganga canal grid in Uttar Pradesh was utilized for pumping of irrigation supplies through 2,200 tube wells in 1948. Similarly, more than 12,500 consumers utilized electricity in Madras state for irrigation pumping. Overall, there was a net increase of 16 million acres (51 million acres to 67 million acres) in the irrigated area during the first plan period.

Nonetheless, for various reasons including finances, technical knowhow, and skilled manpower, the commercial generation of electricity was extremely slow. The average per capita consumption of electricity was only 14 kilowatt-hour per year[3] and in a few states the average consumption was even below one kilowatt per capita per annum. Only five states had per capita consumption higher than the average

---

[3]In comparison, the per capita consumption in the UK was 1100 kilowatt-hour, 2207 kilowatt-hour in the USA, and 3905 kilowatt-hour in Canada.

Original Unit	MTCR	MTCE
One million ton of coal	1	1
One million ton of oil	6.5	2.6
One Terra watt hour	1	0.123
One million ton of firewood (4750 Kcal/Kg)	0.95	0.95
One million ton of dry animal dung	0.40	0.48
One million ton of agricultural waste (4200 Kcal/Kg)	0.95	0.84
Ref – [Planning Commission., 1979]		

Fig. 2.4 Conversion value of different sources of energy to MTCR and MTCE

mentioned above and about 3% of the country's population in six large towns got the benefits of 56% of the total public utility installations. Besides, three sources of commercial energy, energy needs were also met from fuelwood, animal, and plant waste (non-commercial) sources as well. A brief description of these sources is mentioned below:

Commercial Energy—From Coal, Oil, and Electricity

Coal, being the principal source of energy, was used for two purposes. One, directly used by industries, railways, and households; and two, for generation of energy in thermal power plants. For the purpose of calculating the production and consumption of total energy, the energy in different forms is aggregated as either million tons of coal replacement (MTCR) or million tons of coal equivalent (MTCE) as described (Planning Commission 1979) in Fig. 2.4.

Non-commercial Energy—From Fuelwood, Agricultural Waste, and Animal Dung

Substantial quantities of non-commercial fuel were used in industries such as sugar mills, brick kilns, unregistered factories, and household industries. The analysis of consumption pattern of such fuels was generally based on sample surveys with not very high level of accuracy. In Indian context, most of the estimation assumes that annual per capita consumption in this category of energy in the household sector is 0.38 tons of coal replacement in rural areas and 0.40 tons of coal replacement in urban areas (Planning Commission 1979).

An interesting feature of non-commercial energy is the fact that most, if not all, types were available at almost zero monetary cost in rural areas. Poor families were spending time for collection of firewood and dung from forest and other public lands, whereas rich would harvest from their own land. Contrary to this, the share of purchased non-commercial fuels in urban areas was very high with at least 50% animal dung and 75% firewood being purchased by users.

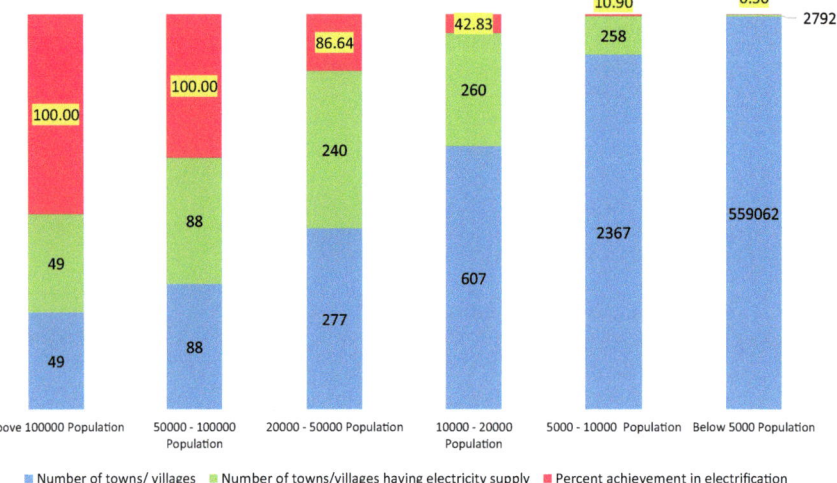

■ Number of towns/ villages   ■ Number of towns/villages having electricity supply   ■ Percent achievement in electrification

Fig. 2.5 State of electrification in India in 1950 (Planning Commission 1953)

Animate Energy—From Bullocks, Camels, Horses, Donkeys

There were 78 million work animals,[4] 12 million animal-drawn carts, and around 42 million ploughshares in 1961. Yet, the estimation of energy derived from such sources has not been backed by authentic scientific data. The animate energy had multiple uses. For example, bullock was used for lifting water, tilling land, and transporting goods and passengers as well. One must appreciate the fact that animate energy has played a significant role in Indian economy. With the gradual increase in technology, the role of animate energy has gradually declined but not written off. One can see millions of animals including, bullock, camel, and donkeys being used for ploughing and transportation.

The necessity of commercial energy for farm irrigation was so acute that other important areas such as semi-urban and rural housing remained bereft of electrification for a long time after independence. The result was that out of approximately 560,000 villages in the country, only a handful (Fig. 2.5) could be connected with commercial energy grid during mid-twentieth century.

It was in 1950 (just before the start of first five-year plan) that the Government of India set up a special commission (known as Planning Commission) with the following objectives:

1. Assess the potential of human (technical or otherwise), material, and capital resources of the country and augment deficiencies if any in terms of national requirements.
2. Formulate plan for most effective and balanced utilization of these resources.
3. Ensure successful implementation of the plan by analyzing and overcoming retarding factors in economic development.

---

[4]Work animals are used for lifting water from wells.

4. Determine the nature of the machinery which will be necessary for securing the successful implementation of each stage of the plan.
5. Appraise from time to time the progress achieved in the execution of each stage of the plan and recommend the adjustments of policy and measures that are necessary.
6. Make such interim or ancillary recommendations as appropriate either for facilitating the discharge of the duties assigned to it.

In the initial plans prepared by the planning commission, the word 'power' was used instead of 'energy' and since power was indispensable for agriculture, the power sector was dovetailed with irrigation sector and emphasis was laid on creation of multipurpose hydropower projects with the objectives of electricity generation as well as irrigation and flood control. This was subsequently changed, and a standalone component of power/energy was introduced.

For the sake of readers convenience and the fact that exhaustive details of progression of power/energy sector is beyond the scope of this book, the rest of this chapter examines the growth of energy sector in India in three segments:

Part I—The initial phase.
Part II—The nuclear energy phase.
Part III—The dawn of renewable energy.

2.2 Part I (The Initial Phase—Coal, Oil, and Water)

2.2.1 The First and Second Five-Year Plans (1951–1956 and 1956–1961)

2.2.1.1 Highlights

(a) Average per capita power consumption was 14 kilowatt-hour/person/year.
(b) Consumption of electricity for agricultural purposes was extremely low before the plan (ranging from 64 million kilowatts in 1939 to 150 million kilowatts in 1949).
(c) Emphasis on improving food production through improved irrigation.
(d) Development of hydroelectric and coal-based thermal power plants dominated the scene.
(e) Diesel-based thermal power plants were also encouraged.
(f) Irrigated area increased by 16 million acres (from 51 to 67 million acres).
(g) Power consumption increased to 25 kilowatt-hour/person/year at the end of first plan.
(h) Both public and private sectors played important role in power generation.

India, at the time of independence, was an underdeveloped country with more than two-thirds of population engaged in low-productivity agriculture, chronic underemployment, diminishing land holdings, decaying cottage, and small-scale industry.

Fig. 2.6 Consumption of
electricity (in million KW)
for agricultural purposes
(*Source* Planning
Commission 1953)

Year	Consumption
1939	64
1945	93
1947	125
1949	150

Power supply at affordable rates was essential for the development of country since the quantum of energy is directly linked with growth and material development index as well as standard of living. Unfortunately, poor access and security of electricity at affordable rates denied commensurate growth of alternative occupation in villages and towns to absorb the growing population. All these factors, coupled with poor education, health, skill, infrastructure, added to their woes and its impact is felt till date. Electrification was an expensive process in those days with a whooping average cost of Rs 60,000–70,000 for providing distribution lines and sub-stations in towns and villages with less than 10,000 population. Figure 2.6 provides the details of electricity consumption for agricultural purposes from 1939 to 1949 (Planning Commission 1953).

There were two broad categories of resources for generating power:

Exhaustible—which included coal, mineral oil, peat, and natural gases. The extractable reserves of coal down to a depth of 1000 feet were estimated to be 20,000 million tons of which good quality coal was only 25% or 5000 million tons. Large reserves of low-grade (i.e. high ash content) coal and lignite were also available for electricity generation. The other known exhaustible resource, petroleum, was scarce with only one source at Digboi, Assam that could fulfil only 5% of country's total demand by producing a meagre 60 million gallons annually.

Inexhaustible—as compared with exhaustible sources, the inexhaustible sources especially water had tremendous potential to meet much of country's power needs for agriculture as well as industrial development. For a large and underdeveloped country like India, the importance of growing energy needs for agriculture and allied operations could never be undermined. It was estimated that water alone could generate 30–40 million kilowatts electricity. For example, in 1948, about 35% of the electric energy generated by the Ganga canal in Uttar Pradesh (about 61 million kilowatt-hour) was utilized for irrigation through 2,200 tube wells. The economy would have been much stronger had we produced enough energy for irrigation, processing of farm produce, cold storage, dairy farming, poultry, and cottage industries at that time. But the problem with hydro projects was the time lag between plan and maturity which was in the range of 5–10 years at that time.

With large population, increased rate of population growth, low farm production, poor technology, poor skill enhancement, and poor resources, India had no option but to divert electricity to farming, irrigation, and allied sector that absorbed about 70% of the workforce and consumed only 4% of electricity in comparison to mining industry that absorbed around 11% workforce but consumed 63% electricity (Fig. 2.7). In contrast, the developed countries like the USA, Germany, UK, and Japan showed

Fig. 2.7 Electricity consumption (million kwh) in 1950 (Planning Commission 1953)

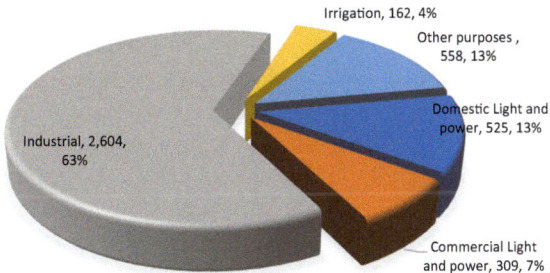

considerable decline in agricultural workforce ranging between 25 and 7% (Planning Commission 1953).

As expected, the first plan focused on the need for energy for growing more food by providing water to the farmland during lean period. The planning commission (Planning Commission 1953) gave priority to agriculture, irrigation, and power sector and provided substantial financial support to these areas out of the proposed expenditure of INR 20,690 million (Fig. 2.8). Keeping in mind the demand for irrigation and power, five new multipurpose hydroelectric projects were proposed at a cost of INR 2000 million. These were:

- Kosi (Stage I);
- Koyna (Stage I);
- Krishna;
- Chambal (Stage I); and
- Rihand.

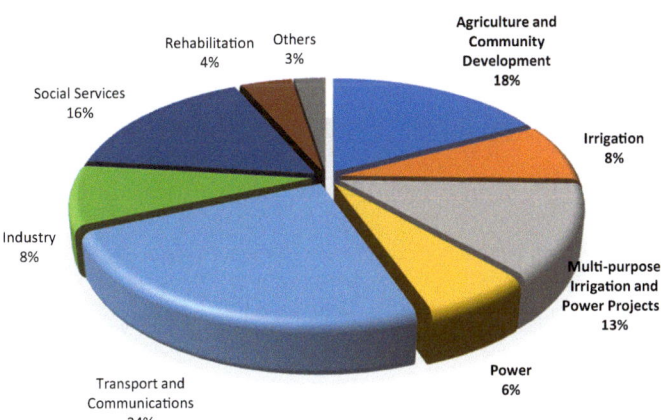

Fig. 2.8 Proposed share of expenditure in different sectors during the first plan (Planning Commission 1953)

Despite high consumption of energy in the industrial sector, there was hardly any reduction in the workforce for agriculture sector as the first plan progressed. The second plan took cognizance of this situation and acknowledged the fact that workforce had to be shifted from agriculture to industrial sector without any relaxation in efforts to increase food productivity. It also recognized that development of power was vital for both agriculture as well as industrial sector, particularly for the development of basic and heavy industries, oil and coal exploration, and development of atomic energy. It emphasized the need for enhancing hydropower potential to 35 million kilowatts of which 20 million kilowatts was expected from the Ganges, Brahmaputra, Indus, and other Himalayan rivers in the northern and north-eastern regions, 7 million kilowatts from the east-flowing rivers of the southern region, 4 million kilowatts from the Narmada, Tapti, Mahanadi, Brahmini, and Baitami basins in the central region, and 4 million kilowatts from the west-flowing rivers.

The installed capacity (Fig. 2.9) which was abysmally poor (2.30 million kilowatts) at the beginning of first five-year plan was more than doubled at the end of second plan (6.9 million kilowatts, including 1 million kilowatts from private sector) (Planning Commission 1956). This was a remarkable achievement since there were several key factors including the capital cost per kilowatt of installed capacity, the foreign exchange component, construction period, and impact of allied activities such as coal mining and washeries. The average cost of production of electricity from hydro, coal, and diesel power station was 1.2, 3, and 25 paisa per unit, respectively. The amount of energy generated increased from about 11,000 million units in 1955–1956 to 22,000 million units in 1960–1961. Forty-four (25 were hydroelectric stations and 19 were thermal stations) hydro and steam power generating projects (new as well as extension to existing ones) were proposed to be undertaken during the second plan.

The consumption of electricity in the industrial and irrigation sector improved manifold during the first and second plan together with other sectors (Fig. 2.10).

	March 1951	March 1956	March 1961
Hydro	0.56	0.96	3.06
Steam	1.00	1.55	2.65
Diesel	0.15	0.21	0.23
Total	1.71	2.72	5.94

Fig. 2.9 Status of installed capacity of power plants (million KW) (Planning Commission 1956)

	1950	1955	1960
DOMESTIC	525	800	1440
COMMERCIAL	309	500	984
PUBLIC LIGHTING	60	110	250
INDUSTRIAL	2609	4600	12000
TRACTION	309	440	655
IRRIGATION	162	260	655
WATER WORKS	182	290	576

Fig. 2.10 Consumption pattern of electricity during first and second plan (million units) (Planning Commission 1956)

Population of Town/village	1950-5l		1955-56		1961
	Total number according to 1941 census	Number electrified as on March 1951	Total number according to 1951 census	Number electrified by March 1956	Number electrified by March, 1961
Over 1,00,000	49	49	73	73	73
50,000 lo 1,00,000	88	88	111	111	111
20,000 to 50,000	277	240	401	366	399
10,000 to 20,000	607	260	856	350	756
5,000 to 10,000	2367	258	3101	1200	1800
Less than 5,000	559062	2792	556565	5300	19861
Total	**562450**	**3687**	**561107**	**7400**	**23030**

Fig. 2.11 Electrification of villages (Planning Commission 1956)

It was also realized during the implementation phase that there was huge gap in demand and supply of electrical energy and there was an acute shortage of power in Delhi, Bombay, and parts of Uttar Pradesh, Tamil Nadu, and West Bengal. In addition, there were around 5.6 lakh small towns and villages in the country at the time of independence of which nearly 5.5 lakhs had population below 5000 individuals. Energizing these towns and villages was an important priority and accordingly 23,000 of them were provided commercial power by 1961 (Fig. 2.11). As a result, the per capita annual consumption of electricity increased from 25 units at the end of the first plan to about 38 units at the end of the second plan (Planning Commission 1956).

For some strange reason, the planner did not highlight the importance of non-commercial energy during the first plan even though fuelwood, animal and plant waste supplied substantial energy to the rural as well as urban household. Nonetheless, the report of the working group on energy policy, prepared by Planning Commission, Government of India in 1979 (Planning Commission 1979) clearly highlights the significance of non-commercial energy. The report mentions that

> Substantial quantities of noncommercial fuels are used in a few major industries such as sugar and in unregistered factories and household industries. It has not been possible to estimate, on a reliable basis, the total contribution of noncommercial fuels to energy consumption in the industries sector. The analysis of consumption of noncommercial fuels in this report relates entirely to their use in the household sector. Even in respect of this sector, there is no record of production and usage of such fuels and it is necessary to estimate consumption from sample surveys…..contd. Briefly, the estimation is based on the assumption that the annual per capita consumption of energy (of all forms) in the household sector is 0.38 tons of coal replacement in rural areas and 0.40 tons of coal replacement in the urban areas.

Between 1953–1954 and 1975–1976, the commercial energy consumption rose from 60.4 to 252.7 million tons of coal replacement, whereas the non-commercial energy consumption figures for the same period was 125.9 to 194.6 million tons of coal replacement. That fuelwood from forests were an indispensable source of household energy (Planning Commission 1953) that can be ascertained from the Indian Forest Statistics of 1947–1948 (Fig. 2.12).

Many power projects were completed during the first five-year plan and that helped in improving energy generation in the country in a big way (Fig. 2.14). The details of 11 power projects that were operationalized by the end of first plan are provided in Fig. 2.13.

State	Production of fuelwood ('000 tons) in 1947-48
Assam	148
Bihar	114
Bombay	985
Madhya Pradesh	1061
Madras	690
Orissa	96
Punjab	335
Uttar Pradesh	1229
West Bengal	302
Hyderabad	107
Jammu and Kashmir	179
Mysore	308
PEPSU	5
Rajasthan	NA
Saurashtra	36
Travancore-Cochin	131
Ajmer	6
Bhopal	10
Bilaspur	NA
Coorg	7
Himachal Pradesh	16
Kutch	NA
Tripura	2
Vindhya Pradesh	8
A&N island	2
TOTAL	5777
Source: [Plan. Comm 1953]	

Fig. 2.12 State wise data of fuelwood production

Nangal (Punjab)	48,000 kW
Bokaro (Bihar)	150,000 kW
Chola (Kalyan, Bombay)	54,000 kW
Khaperkheda (Madhya Pradesh)	30,000 kW
Moyar (Madras)	36,000 kW
Madras City Plant Extensions (Madras)	30,000 kW
Machkund (Andhra and Orissa)	34,000 kW
Pathri (Uttar Pradesh)	13,600 kW
Sarda (Uttar Pradesh)	27,600 kW
Sengulam (Travancore-Cochin)	48,000 kW
Joe (Mysore)	72,000 kW

Fig. 2.13 Completed power projects by the end of first plan (Planning Commission 1953)

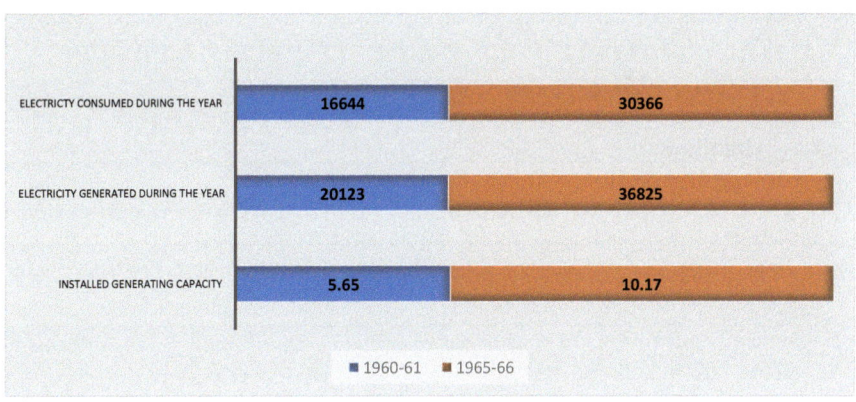

Fig. 2.14 Growth of electricity generation between 2nd and 3rd plan (million kilowatts) (*Source* Planning Commission 1970)

As far as growth in the share of commercial energy is concerned, it is mentioned that the decade of 1950s witnessed an upsurge in the transport sector and substantial downslide in industrial sector indicating growth reduction (Fig. 2.15).

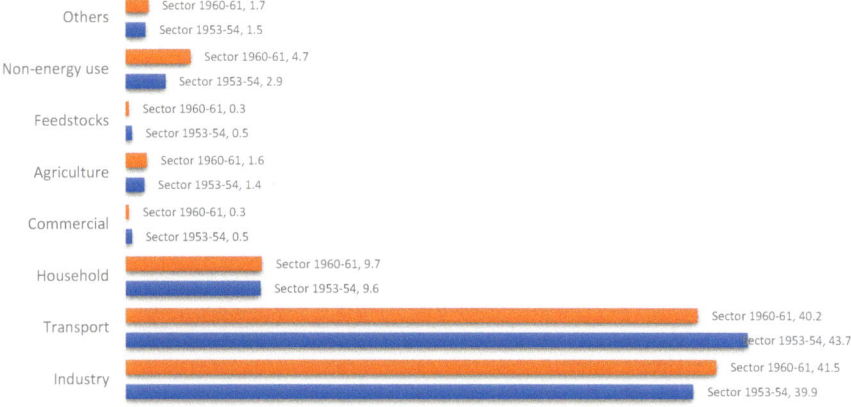

Fig. 2.15 Percentage share in final commercial energy consumption by different sectors (*Source* Planning Commission 1970)

2.3 Part II—The Nuclear Energy Phase

2.3.1 The Third (1961–1966), Fourth (1969–1974), and Fifth (1974–1978) Five-Year Plans

2.3.1.1 Highlights

(a) Focus on development of efficient small-scale industries in towns and rural areas for improving employment opportunities and living standards.

(b) Reduced efficiency of thermal power stations due to aging process (the overall efficiency of thermal power stations in 1959–1960 was only 19.5%).

(c) Crude oil and natural gas production were abysmally low, being 8.4 million tons and 2.4 billion cubic meters in 1975–1976, respectively.

(d) 13.6 million tons of crude was imported in 1975–1976.

(e) Focus of rural electrification continued for energizing tube wells and pump sets. An ambitious target of energizing 81,000 additional villages and 1,300,000 additional pump sets was set.

(f) Decline in fuelwood availability and sensitivity toward environmental issues attracted public attention.

(g) New primary sources of energy were identified for power generation in addition to the old ones.

    i. Coal—the total reserves of coal was estimated to be 50,000 million tons.

    ii. Lignite—the estimated reserve was estimated to be about 2000 million tons.

    iii. Hydropower.

    iv. Uranium and thorium.

    v. Oil and natural gas.

    vi. Tidal power, wind power, geothermal power, and solar radiations.

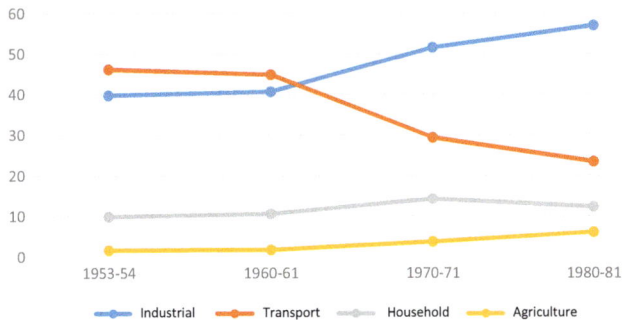

Fig. 2.16 Percentage share of energy consumption in different sectors (*Reference* Planning Commission 1992)

(h) Tarapur power station was commissioned with two reactors of 150 megawatts each.
(i) Population growth rate at 2.5% per annum was disproportionately high in proportion to available resources, low income, and poor economic growth.
(j) A multipronged approach to develop biogas technology, as well as new sources such as solar energy, tidal, and wind power, was initiated.
(k) Consumption of agricultural wastes for fuel in 1975–1976 was estimated to be around 41 million tons.
(l) The number of bullock carts increased from 12 to 13 million during the 1960s clearly signifying the importance of animal energy in Indian economy.

Burgeoning population and consequent food requirement forced the planners to increase the irrigation potential in every plan and the same finds reflection in the quantum of area irrigated year on year. At the beginning of the first plan, the area irrigated was 51.5 million acres and the third plan envisaged an increase of 125 million acres from major and medium irrigation projects as well as from tanks, tube wells, and open wells. Besides increasing irrigation potential, the third plan also aimed at developing efficient small-scale industries in small towns and in rural areas to increase employment opportunities, raise incomes and living standards, and bring about a more balanced and diversified rural economy. However, availability of electricity was a major limiting factor in achieving this objective. Share of energy consumption in transport sector declined, and household as well as agriculture sector (Planning Commission 1992) did not reflect appreciable growth (Fig. 2.16). Solar, tidal, geothermal, and wind as resources for power were still a distant dream. However, some research work was carried out to utilize wind power, but high per unit cost for building wind power stations was an impediment to extensive introduction of such plants for power generation. As far as non-commercial energy was concerned, the third plan acknowledged the fact that due to severe shortage of fuelwood,[5] almost 400 million tons of cow dung (=60 million tons of fuelwood) was burnt as cooking fuel.

---

[5]The third plan also anticipated a shortfall of 100 million tons of fuelwood by 1975.

Unfortunately, implementation of third five-year plan suffered huge setbacks. The outbreak of hostilities in 1962 and 1965 and other factors delayed the implementation of many projects including those of power generation that had a shortfall of 2.52 million kilowatt. The actual capacity commissioned was 10.17 million kilowatts against the target of 12.69 million kilowatts. Rural electrification was worst hit with only 20,000 villages electrified, most of them having population less than 5000.

There was a gap of three years between third and fourth plan during which period three annual plans were prepared and implemented. Priority was accorded for the completion of old projects and about 4.12 million kilowatts of generating capacity were added between 1966 and 1969. By the end of March 1969 some 71,280 villages were electrified and 1,087,567 irrigation pumps were energized for improving farm production.

The year 1974 was a defining moment and will be remembered as the year of 'black outs' and industrial shutdowns. In fact, India had made great strides by enhancing its energy requirement from 30 to 54% through commercial energy (oil, coal, hydro, thermal power, and nuclear) between 1954 and 1974. But a sudden and sharp increase in the price of crude oil by OPEC in 1974 made the cost of imported oil prohibitive for India. India's bill for crude and petroleum products went up from Rs. 2040 million in 1972–1973 to Rs. 5600 million in 1973–1974 and thereafter to Rs. 11,300 million in 1974–1975 (the price of crude went up from USD 3 per barrel to USD 14 per barrel). This was exacerbated by failure of monsoon and inadequate coal availability. Food insecurity compelled the government to divert energy to agriculture sector resulting in massive shortfall to the industry. India topped the UN list of 30 developing countries that were most seriously affected by the global energy crisis. Such a massive increase in oil prices in three years forced India to cut down its oil consumption from the estimated 25 million tons in 1974–1975 to 21.5 million tons. Almost 42% of all foreign exchange earned by the country from its exports had to be used to pay for its oil and oil product imports. India had no option but to increase indigenous coal production and the fifth plan envisaged an increase of 57 million tons in coal production by 1978–1979 from the 78 million tons production in 1973–1974.

Anticipating India's energy insecurity in mind, the scientists and planners in India had started its nuclear energy program in 1948 with the establishment of Atomic Energy Commission. The first atomic energy establishment set up in Trombay in 1954 processed thorium uranium cake and produced thorium nitrate and uranium whose fuel value was equivalent to 10,000 tons of coal per one ton of uranium (Planning Commission 1961). Subsequently, other plants/projects such as extraction of uranium ore, uranium purification, heavy water, and atomically pure graphite production were undertaken as well. Since India had one of the largest deposits of thorium in the world, the nuclear energy program was divided into three phases, viz.:

Stage 1—Utilize natural uranium as fuel for power generation and production of plutonium.
Stage 2—Install reactors for using plutonium as fuel and convert thorium to uranium-233.
Stage 3—Use uranium-233 with thorium in breeder reactors.

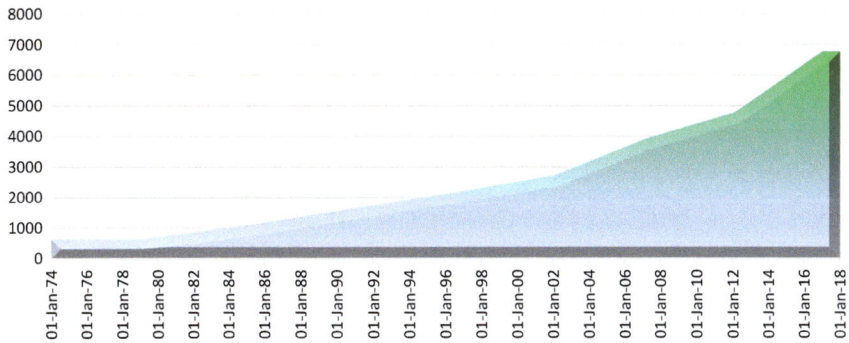

Fig. 2.17 Growth of nuclear power generation in India (installed capacity in MW) (*Source* CEA 2018)

India's first atomic power station at Tarapur with 380-megawatt capacity went into operation in 1969. Thereafter, two more units were operationalized at Rana Pratap Sagar Station in Rajasthan—the first unit of 200 megawatts was commissioned in 1973 and the second one in 1974. This was followed by setting up of a third and fourth nuclear power plant at Kalpakam in Tamil Nadu and Narora, UP.

The expansion of nuclear power (Fig. 2.17) received setback after the first nuclear explosion (in fact implosion) was carried out in 1974. Nuclear power production almost stagnated for a decade and picked up gradually thereafter due to severe restrictions imposed by nuclear supplier group including on supply of equipment and transfer of technology. As a result, the nuclear energy program was severely crippled and once again India was forced to rely on coal and oil.

The fifth five-year plan continued the program of expansion of irrigated area and rural electrification in the country and aimed at energizing 13 lakhs pump sets and electrification of 81,000 villages (Fig. 2.18). By 1979–1980 substantial progress was attained and almost 2.5 lakh villages were electrified (Planning Commission 1980).

At the same time, India continued to rely heavily on fuelwood for household cooking and small industries so much that it overshot the supply from forests in 1978–1979 when the demand went up to 132 million tons (and only 94 million tons fuelwood was supplied from forest). The excess demand was met by supplementing

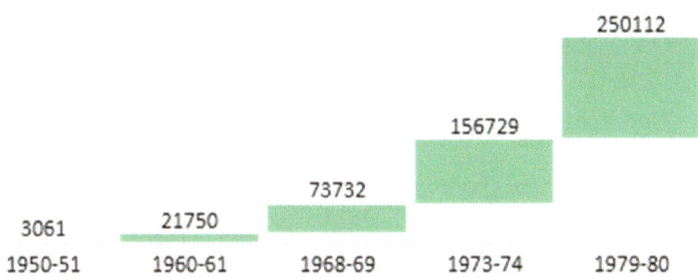

Fig. 2.18 Growth in electrification of villages in three decades after independence (*Reference* Planning Commission 1980)

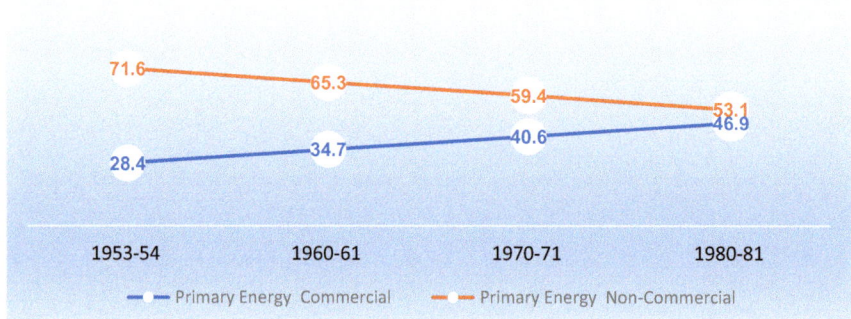

Fig. 2.19 Changes in pattern of commercial and non commercial energy consumption (%) (*Reference* Planning Commission 1992)

agricultural waste and animal dung. Analysis of energy trend over three decades reveals that the process of replacement of non-commercial energy to commercial energy was extremely slow as can be seen from Fig. 2.19 (Planning Commission 1992).

In fact, India's per capita consumption (Fig. 2.20) of commercial energy has remained abysmally poor from the independence year itself. From 14 units per capita per year, we have achieved 718 units per capita per year, an increase of 704 units in 68 years. In other words, only 10 units per capita have been added annually. Factors like electricity generation, transmission losses, theft, cost per unit, human population, access (rural areas in particular), and paying capacity have played critical role in poor per capita consumption. At the current rate, India's per capita annual consumption is expected to be around 2900 units in 2040. This assessment, if correct, is quite disheartening as we will be marginally above 1951 consumption pattern of USA and much below that of Canada (in 1951, per capita annual consumption was 2207 units and 3905 units in USA and Canada, respectively[6]).

2.4 Part III (Emergence of Renewable Sources)

2.4.1 The Sixth Five-Year Plan (1980–1985)

2.4.1.1 Highlights

(a) Country was divided (figuratively) into two groups:

    i. Household dependent on non-commercial energy and
    ii. Industries dependent on commercial energy

---

[6]Reference—First Five-Year Plan. Planning Commission, Government of India.

YEAR	CONSUMPTION (KwH)
1951	14
1956	25
1961	38
1966	61.4
1968-69	79
1973-74	97.5
1977-78	120.7
1984-85	168.5
1985-86	178
1986-87	191
1987-88	201
2005	380*
2009	503*
2012-13	914.41
2013-14	956.64
2014-15	1010
2015-16	1075 (765*)
2016-17	1122 (805*)
December 2018	718.3#
2040 (Anticipated)	2911 – 2924@
Source – Compiled by author from 1 [Plan. Comm.,1953]; [Plan. Comm.,1956]; [Plan. Comm.,1961]; [Plan. Comm.,1970]. 2 [Plan. Comm. Stat. Profile, 2001]. *3 [World Bank. Little Green Data Book 2017. World Bank 2017]. # 4 [Sabha, Lok. "Lok Sabha Question 337 (Archive)" (2019)] @ [Aayog, 2017]	

Fig. 2.20 Per capita consumption (kilowatt hour)

(b) Per capita energy consumption continued to be dismal, 10% of global average, which was a poor reflection of household income.
(c) High dependency on petroleum for commercial energy.
(d) Fuelwood contributed almost 75% of the total energy but was becoming scarce.
(e) Increase in international crude prices.

India continued to rely heavily on non-commercial energy sources (40–45% of the total consumption) till early 1980s with more than 90% of the total rural households using fuelwood, animal, and plant waste for cooking while kerosene was used by only 5%. Around 80 million animals provided draught energy for farming and transport. It is noteworthy to mention that the concept of village electrification was misleading and exaggerated. Even though 0.25 million villages (out of a total of 0.57 million) were electrified on record but the fact that only 14% of the total households were electrified exposed the dichotomous approach adopted by the government in defining 'village electrification' and 'household electrification'. Moreover, household electrification was limited to lighting and fans and certainly not for cooking purpose.

While rural households relied on non-commercial sources, the industrial sector was the major beneficiary of commercial energy (around 38.5%) with rising energy intensity. The electricity consumed by industry per rupee (of value addition) went up from 0.54 kilowatt-hour in 1960–1961 to 1.02 kilowatt-hour in 1975–1976 with relatively high proportion of power-intensive metal industries that contributed to the high electricity intensity.[7] Spurt in industrialization and economic intensification led to increased consumption of petroleum products also despite three-fold rise in crude oil prices. During the period 1978–1980, oil exploration, both offshore and onshore, continued to receive high priority. Onshore exploratory activities in Assam-Arakan region were stepped up and offshore seismic surveys were conducted over a large part of the continental shelf. Exploratory drilling was done in the Bombay offshore area and off the Godavari basin, Kerala coast and Mangalore coast.

Out of three primary sources, only two, that is coal and water, were indigenous and most dependable but rapidly and disproportionately growing population brought down their per capita availability. For example, India had coal reserves of 176 tons per person as compared to 1168 tons/person in China, 22,066 tons/person in USSR, and 13,448 tons/person in USA. The proven reserves of oil were also very limited, being only 0.55 tons per capita as against 34.83 tons in USSR, 16.32 tons in USA, and 2.86 tons in China. On nuclear energy front also, the guaranteed uranium resources in India were about 34,000 tons of which about 15,000 tons were economically exploitable and enough for the development of first-stage nuclear power program of about 8000 megawatts of installed capacity. In addition to uranium, India was endowed with 363,000 tons of thorium deposits for sustaining for subsequent stages of nuclear program.

In order to maintain annual economic growth rate of 5.5–6% India required 426 million tons coal, 69 million tons oil, and 464 TWh (terrawatt-hour) electricity generation by the turn of the twentieth century. One of the reasons for huge energy demand was abnormally high loss (20%) of energy in the production, transmission, and distribution of electricity itself. To add to the woes, 60% of oil requirement was met through imports consuming two-thirds of foreign exchange reserves annually.

Overall, grim scenario of commercial as well as non-commercial energy sources led to the emergence of renewable energy in the form of solar, wind, biogas, and energy plantation. For a tropical country like India, solar energy was an abundant source of renewable energy. However, technical wherewithal for concentrating solar energy for high heat applications was expensive and efforts were initiated to bring down the cost of solar-grade silicon material and improve the efficiency of solar cells. Considerable progress was achieved in the development of solar water heaters, solar cookers, solar driers, and solar pumps. At the same time, biogas plants were introduced at the rural and peri urban level to convert animal waste into energy. A new program for fuel and farm forestry was taken up with the target of 13 lakh hectares to minimize ecological damage due to excessive harvest of fuelwood from the forest areas. The broad approaches for the development of renewable sources of energy in the sixth plan were:

---

[7]Commercial energy consumed per rupee of product.

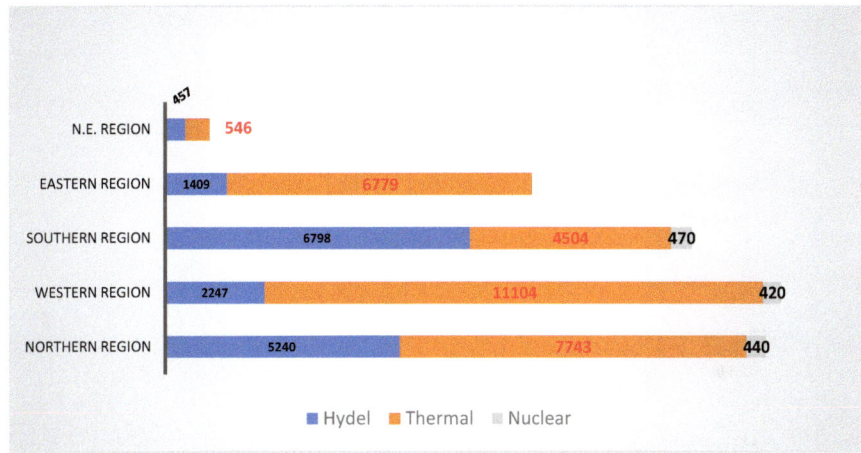

Fig. 2.21 Anticipated growth in commercial energy generating capacity by the end of sixth plan (MW) (*Reference* Planning Commission 1980)

i.   Field testing and demonstration of commercially viable technologies.
ii.   Intensification of research and development of alternative technologies.
iii. Large-scale programs of energy forestry and biogas.

Considering all limitations, the energy strategy for the sixth plan focused on faster exploitation of domestic energy resources by accelerating exploration of oil and gas as well as rapid development of proven hydrocarbon reserves. Despite long gestation period and cost involved in developing hydroelectric power, the planners pushed for the same in north-east and north-west regions of the country. It was also decided that in order to be self-reliant in nuclear power technology, construction of conventional nuclear reactors should continue, and abundant thorium reserves should be exploited for future breeder reactors. Figures 2.21 and 2.22 provide the details of region-wise expected growth in commercial energy and state-wise installed capacity, respectively, at the end of sixth plan (Planning Commission 1980).

2.4.2   The Seventh Five-Year Plan (1985–1990)

2.4.2.1   Highlights

(a) The proportion of commercial and non-commercial energy reached almost 50:50 level.
(b) Indiscriminate use of non-commercial energy sources led to energy crisis in the rural areas.
(c) Only 50 million tons of fuelwood were available per annum against the demand of more than 150 million tons.

States	Installed Capacity (MW)
Andhra Pradesh	3025.43
Assam	569.78
Bihar	1615.27
Gujarat	3396.02
Haryana	1541.21
Himachal Pradesh	270.83
J&K	206.18
Karnataka	2529.8
Kerala	1136.5
M.P.	3033.02
Maharashtra	6196.3
Manipur	10.41
Meghalaya	131.11
Nagaland	4.68
Orissa	1483.12
Punjab	2216.84
Rajasthan	1315.6
Sikkim	15.04
Tamil Nadu	2959
Tripura	19.06
Uttar Pradesh	5311.76
West Bengal	2978.54

Fig. 2.22 Cumulative installed capacity at the end of sixth plan (as on March 1985) (Reference Planning Commission 1980)

(d)   Out of 324 million tons of animal dung (air-dry) produced per year about 73 million tons were burnt for energy purposes.

(e)   Per capita consumption of energy (commercial) was only 1/8 of the global average.

(f)   New estimate of uranium resources in the country was assessed at 70,000 tons.

(g)   High cost of harnessing solar energy prevented introduction of solar energy at commercial scale.

Despite increase in commercial energy production, the rural areas continued to suffer electrification deficiency especially household connections. Eighty percent of commercial energy was consumed by 24% urban population and though 64% villages were electrified yet only 8% households were provided electricity. Keeping the deteriorating commercial energy situation in mind, the energy strategy was modified in the seventh plan with the following main elements:

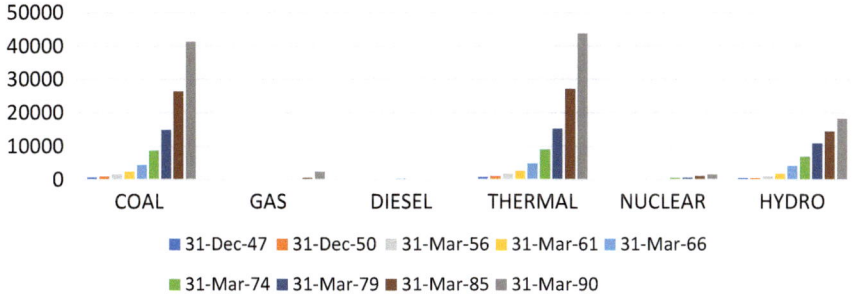

Fig. 2.23 Growth in installed capacity (MW) of commercial energy (1947–1990) (*Source* Energy Statistics 2018)

i.   Continued and accelerated exploitation of coal, hydro, and nuclear power (Fig. 2.23).
ii. Intensified exploration for oil and gas, and exploitation of oil.
iii. Conservation of energy.
iv. Exploitation of renewable sources of energy like energy forestry, biogas, biomass, wind, solar energy, and so on, especially to meet the energy requirements of rural communities.

2.4.3 The Eighth Five-Year Plan (1992–1997)

2.4.3.1 Highlights

(a) The share of non-commercial fuel declined from 74% in 1950–1951 to 41% in 1991.
(b) Rural household sector continued to meet its energy from non-commercial sources in a sizeable proportion with fuelwood accounting for 65% and the balance coming from agriculture and animal waste.
(c) Country's wind energy potential was estimated to be 10,000 megawatts.
(d) The efficiency at which useful commercial energy was realized from resources was extremely poor (between 10 and 15%).
(e) Fuelwood, agriculture, and animal waste continued to be significant for Indian household (Fig. 2.24).

One of the adverse consequences of increased energy consumption in the urban areas was greater dependence on import of crude oil and petroleum products that caused heavy drain on the foreign exchange reserves of the country. Oil prices, after a fall in the late 1980s, showed a steep upward trend during the Gulf War. Taking into account the prospects and potential of different onshore and offshore basins in the country, it was highly unlikely that indigenous production could be augmented. Against this backdrop, a long-term energy strategy was developed that accorded highest priority to renewable energy with increasing emphasis on demand management, conservation, and efficiency.

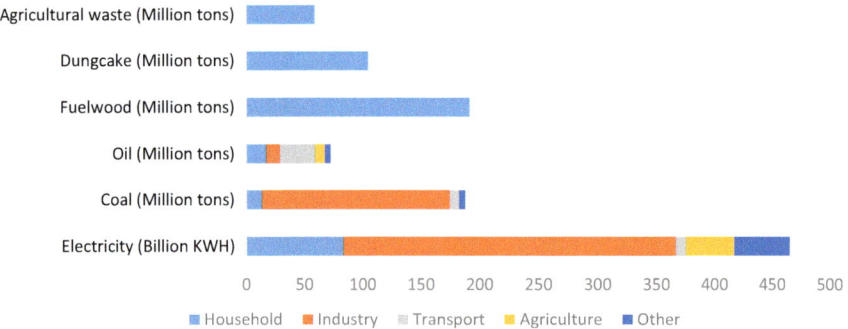

Fig. 2.24 Commercial and non-commercial energy requirements in 1999–2000 (*Source* Planning Commission 1992)

The eighth plan suggested the following short-term, medium-term, and long-term priorities:

Short-Term

i.   Maximization of returns from existing assets in the energy sector.
ii. Reduction of technical losses in production, transportation, and end-use of all forms of energy.
iii. Reduction in energy intensity of the different energy-consuming sectors and energy conservation.
iv. Meeting basic energy needs of the rural and the urban households, and to reduce the existing inequities in energy use.
v.   Maximization of demand for energy from indigenous resources.

Medium-Term

i.   Substitution of petroleum products by coal, lignite, natural gas, and electricity.
ii. Accelerated development of all renewable energy resources, especially the available hydroelectric potential.
iii. Research and development on decentralized energy technologies based on renewable resources.

Long-Term

i.   Promotion of energy supply system based largely on renewable sources of energy.
ii. Promotion of technologies of production, transportation, and end-use of energy that are environmentally-benign and cost-efficient.

While total energy supplies (both commercial and non-commercial) increased from 82.7 MTOE (million tons of oil equivalent) in 1950–1951 to about 291 MTOE in 1990–1991, the share of non-commercial fuel continued to be substantial, that is, 41% in 1990–1991. Fuelwood alone accounted for 65% of the total non-commercial energy consumed in the country. Among the indigenously produced primary commercial fuels, the relative share of oil and gas increased from 1.2% in 1950–1951 to 33% in 1990–1991, whereas the share of coal declined from about 98% in 1950–1951 to 61.8% in 1990–1991. More and more commercial energy was supplied to the industrial sector at the expense of reduction in transport sector. One positive factor was marginal improvement in commercial energy share in household sector after 40 years of independence (Fig. 2.25).

During the same period, allocation of financial resources increased from almost 20 to 28% during the first 40 years after independence. The rise was not substantial considering the energy requirement of the country (Fig. 2.26).

A noticeable feature of the eighth plan was push for renewable energy with focus on rural areas. Solar cookers, improved cookstoves, biogas (particularly based on

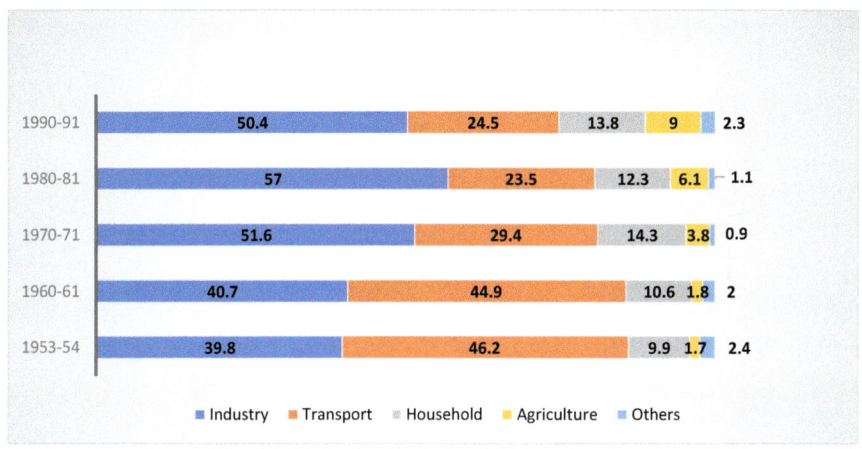

Fig. 2.25 Decadal growth (% share) of different sectors in commercial energy consumption (*Source* Planning Commission 1992)

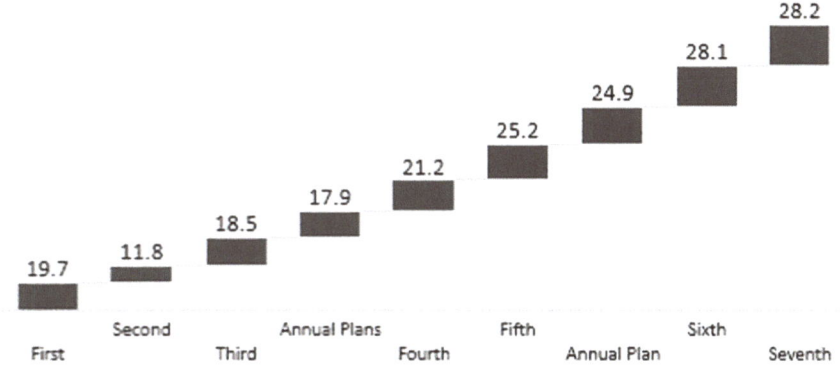

Fig. 2.26 Percent share of energy sector in plan outlay (*Source* Planning Commission 1992)

	Target	Achievement
Biogas Plants	7.5 lakh	9.60 lakh
Improved Chulhas	100 lakh	127.00 lakh
Solar cookers	3.0 lakh	1.98 lakh
Wind Power Generation	100 MW	860 MW
Biomass Power and Co-generation	300 MW	115 MW
Small Hydro	200 MW	93 MW

Fig. 2.27 Achievements in renewable energy sector during eighth plan (*Source* Planning Commission 1992)

animal waste) were introduced in a big way (Fig. 2.27) with the purpose of revolutionizing rural energy landscape. It did succeed in the initial phase but failed to sustain for various reasons, including maintenance, food preferences and taste, unwillingness to use gas generated by animal waste (gobar) for cooking food, and so on.

2.4.4   The Ninth Five-Year Plan (1997–2002)

2.4.4.1   Highlights

(a) Fuelwood continued to be the backbone of rural cooking and heating.
(b) Use of wood (including fuelwood) exceeded the annual increment resulting in accelerated deforestation and environmental degradation.
(c) Per capita commercial energy consumption continued to remain low.
(d) India remained energy deficient with heavy reliance on import of coal and petroleum.
(e) The annual availability of wet animal dung for burning was about 960 million tons and this consequently reduced the availability of dung as a valuable organic manure.
(f) Total availability of crop residues for fuel was estimated to be 450–500 million tons.
(g) The draught animal population in the country was estimated at 70 million and animal energy continued to be used in agricultural operations and for rural transportation.
(h) The share of commercial energy in total primary energy supply increased from 28% in 1950s to 66% in 1996–1997.

(i) The share of non-commercial fuels declined from 74% in 1950–1951 to about 34% in 1996–1997.

(j) In 1996–1997, India imported crude oil and petroleum products worth $9.3 billion (Rs. 31,000 crore).

During the ninth five-year plan, the term renewable energy was replaced by non-conventional energy and for the first time a pan India assessment of renewable energy potential was carried out and the same is depicted in Fig. 2.28.

The assessment indicated tremendous scope from coastal sources such as ocean thermal, sea wave, and tidal power but so far actual production has remained elusive. At the same time, assessment of cookstoves and biogas plants also indicated huge demand but again success was limited (Fig. 2.29).

The ninth plan also carried out assessment of substitution of non-commercial energy by commercial over a period of several decades (Fig. 2.30). Since the proportion of commercial energy was increasing with every plan period and the fact that India was importing huge quantities of oil at continually increasing prohibitive cost (Fig. 2.31), the issue of energy security became a matter of concern. In view of

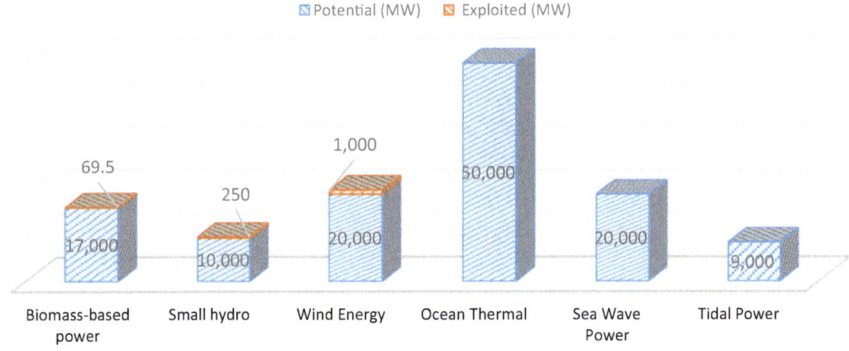

Fig. 2.28 Renewable energy potential (Source Planning Commission 1997)

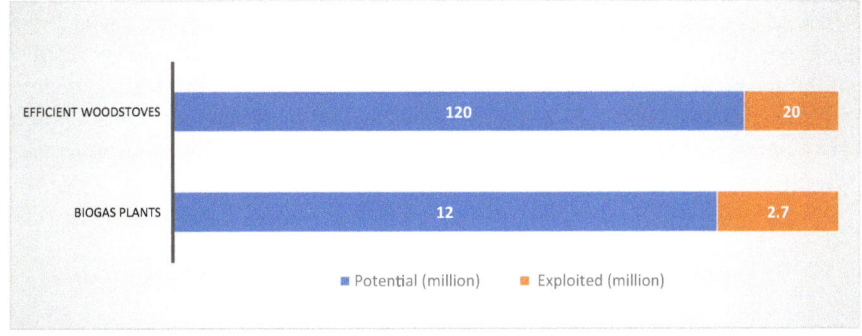

Fig. 2.29 Potential for efficient cookstoves and biogas plants (*Source* Planning Commission 1997)

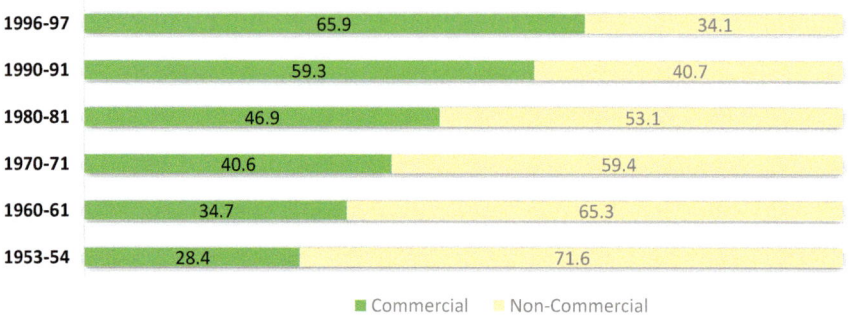

Fig. 2.30 Percent change in pattern of commercial and non commercial energy consumption (*Source* Planning Commission 1997)

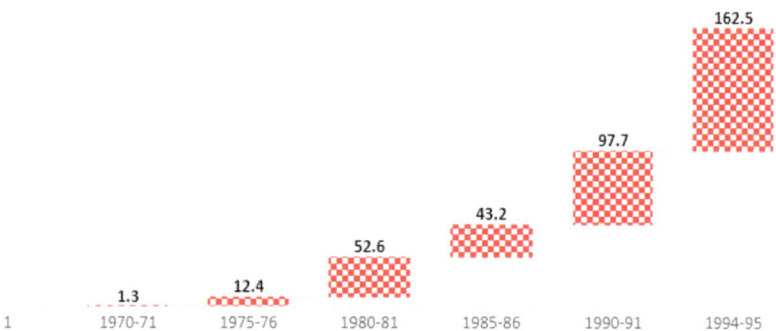

Fig. 2.31 Rise in oil import cost (billion INR) (*Source* Planning Commission 1997)

stagnant domestic oil production, higher imports and uncertainties regarding prices and availability made the developing countries like India more vulnerable than the developed countries. To add to the woes, high ash content of Indian coal generated high quantities of fly ash leading to detrimental effect on environment, and thus the introduction of clean technologies became essential.

The key issues regarding energy implications that India faced were rising population, need for economic growth, access to adequate commercial energy supplies and the financial resources needed to achieve this, rational energy pricing regime, improvements in energy efficiency of both the energy supply and consumption, technological upgradation, a matching R&D base, and environmental protection.

2.5 The Complexities of Renewable Energy

Almost five decades after independence, the poor household sector in both urban and rural areas continued their dependency on fuelwood as a result of which the total requirement of fuelwood stood at 200 million tons per annum of which 102 million tons was to be obtained from forest areas and the rest from farm forestry. To overcome the problems associated with fuelwood, biomass energy programs comprising biomass production, biogas, improved cookstoves, gasifiers were encouraged and implemented in the ninth plan with the aim of generating at least 500 megawatts.

Solar energy has its complexities and one of the major impediments that prevented widespread application of solar energy was its cost. Compared with other conventional sources, the cost of solar photovoltaic cells, their installation, and production of electricity was abnormally high. As a result, the use of centralized solar energy system was deferred and decentralized applications such as lighting, water pumping, water heating, and cooking were encouraged and stand-alone systems were installed for applications like the solar thermal water heaters, solar cookers, solar dryers, solar desalination systems and solar photovoltaic street lights, solar photovoltaic domestic lights, and solar lanterns. Besides solar energy, technological options to produce electricity from waste including from industrial effluents, municipal/urban waste, tannery waste, vegetable/market yard waste, sewage, pulp, and paper industry waste were also explored for the future.

As the energy challenges were getting more and more complex with social, environmental, economic, and security issues, the planners decided to adopt certain measures to reduce the energy intensity of the economy. These included.

i. Demand management through greater conservation of energy, optimum fuel mix, structural changes in the economy, an appropriate modal mix in the transport sector, that is greater dependence on rail than on road for the movement of goods and passengers.
ii. Measures to improve the transport infrastructure, viz., roads, better design of vehicles, use of compressed natural gas (CNG) and synthetic fuel.
iii. Move away from depletable to inexhaustible resources, viz., solar, wind, biomass energy.
iv. Greater emphasis laid on the exploitation of hydroelectric power, particularly for meeting peak demand.

2.5.1 The Tenth Five-Year Plan (2002–2007)

2.5.1.1 Highlights

(a) More than 60% households depended on traditional sources of non-commercial energy, such as fuelwood, dung, and crop residues.
(b) Fuelwood consumption was 130 million tons in 2001–2002 which constituted more than 65% of non-commercial energy source in rural household.

(c) Animal dung and plant waste consumption during 2001–2002 was 130 million tons.

(d) The share of hydrocarbons in the total energy consumption increased from 37.2% in 1980–1981 to 44.9% in 2001–2002.

(e) Share of non-commercial energy in the total primary energy supply declined from 53.1% in 1980–1981 to 31.8% in 2001–2002.

Twenty-first century witnessed increased production and consumption of coal worldwide (despite clear signs of earth warming by coal and oil) with China (1171 MT), USA (899 MT), and India (310 MT) leading the production (Planning Commission 2002a, b). Together, coal and natural gas contributed to around 22% of global primary energy, each against 40% from oil, 7% from nuclear, 2% from hydro, and 6% from renewables. Being abundant (available in 100 countries[8]) and responsible for 38% global electricity generation, coal was expected to continue as an important source of global energy at least in coming decades. For India, coal was and will remain indispensable (Box 2) as it shared 50% of primary commercial energy supply and it is too early to predict a switch over to other sources. Of the 104,917 megawatt overall installed power generation capacity in the country (as on March 31, 2002), about 59,386 megawatts were coal-based and 2745 megawatt was lignite-based, totaling to 62,131 megawatts or 59%. In fact, the tenth plan anticipated that even with an estimated production of 405 million tons in 2007, there would be a shortfall of more than 55 million tons.

Not only coal, the demand for oil and natural gas also was likely to increase primarily due to the expansion of transportation sector. The domestic production of crude oil was only 32.03 million tons in 2001–2002 with upward trajectory to reach 33.97 million tons in 2006–2007. At the same time, the demand for petroleum products was assessed to grow from 99.13 million tons in 2001–2002 to 134.6 million tons in 2007. Similarly, India's natural gas production was expected to reach 37.62 BCM in 2007 from the existing 29.69 BCM in 2001–2002.

---

[8]Proven coal reserves have been estimated to last for over 200 years. In contrast, proven oil and gas reserves are estimated to last around 40 and 60 years, respectively, at current production level in 2001–2002. The geological coal reserves in India were estimated at 220.98 billion tons as on January 2001 with proven reserves of 84 billion tons.

Box 2 - The Indispensable Coal **[Source - Plan. Comm., 2002]**

1. In 1999, coal was responsible for 38% of total global electricity generation.

Country	Coal's share in electricity generation
India	70 percent
USA	56 percent
China	80 percent
Australia	84 percent
South Africa	90 percent
Germany	51 percent
Poland	96 percent

2. Coal based thermal power generation has a shorter gestation period and low investment cost when compared with other commercial energy resources like nuclear or hydropower.
3. Coal prices remained more stable as compared to the volatility in oil and natural gas prices.
4. Coal is safe and easy to transport.
5. The efficiency of energy conversion can reach up to 45% through the use of supercritical steam.
6. The negative aspect (environment pollution) can be mitigated by clean technologies as they can substantially reduce the carbon di oxide emission per unit of energy produced. A 5% conversion efficiency can bring more than 10% reduction in carbon di oxide emission

Eventually, the energy pathways for India pointed toward reliance on hydrocarbon fuel and accordingly a hydrocarbon vision 2025 was prepared with the following priorities:

Natural Gas

1. Natural gas was identified as a preferred option for future.
2. Creation of facilities for import of LNG and setting up of pipelines from major gas-producing countries.

Oil Security

1. Acceleration of exploration efforts, especially in deep offshore and frontier areas.
2. Improved oil recovery.
3. Equity in oil and gas abroad.
4. Strategic storage facility for crude oil.

Alternative Fuels

1. Development of alternative fuels such as coalbed methane, ethanol blending, gas hydrates, and fuel cells.
   Coalbed Methane
   It was estimated that India had around 1000 billion cubic meters of coalbed methane which was likely to emerge as a new source of commercial energy.
   Blending of Petrol and Diesel with Ethanol
   A decision was taken to blend petrol and diesel with 5% ethanol to be obtained from sugar industry.
   Gas Hydrates
   Undertake drilling operations to produce gas from gas hydrates.

Alongside hydrocarbon vision, the focus on nuclear energy was also reinforced and a three-stage approach was envisioned:

Stage 1: Setting up 10,000-megawatt plant based on pressurized heavy water reactor using indigenous natural uranium resources.
Stage 2: Introduce fast breeder reactor technology using plutonium extracted from reprocessing of spent fuel from stage 1.
Stage 3: Utilize indigenous and natural thorium resources for power generation.

Since India is poorly endowed with natural uranium, the three-stage program was aimed at converting thorium to fissile material. The available uranium resources support only 10,000-megawatt power generation. Nonetheless, the potential of natural uranium could be enhanced to 30,000 megawatt through fast breeder reactors that use plutonium obtained from recycled spent fuel of stage 1.

Renewable sources of energy, especially wind, small hydro, and biomass, were gaining grounds (Fig. 2.32) in the overall scheme of power generation (Planning Commission 2002a, b). But renewables were nowhere close to non-renewables.

Fig. 2.32 Target and achievement of renewable energy (*Source* Planning Commission 2002a, b)

	10th Plan Target (MW)	10th Plan Achievement (MW)
Wind	1500	5415
Small Hydro	600	520
Waste to Energy	80	25
Solar Photovoltaic	5	1
Solar Thermal Power	140	0
Biomass	700	750

2.5.2 The Eleventh Five-Year Plan (2007–2012)

2.5.2.1 Highlights

(a) Over half of the country's population was devoid of electricity or any other form of commercial energy.

(b) More than two-thirds households in the country were using non-commercial fuels for cooking.

(c) As such there was no statistical assessment of non-commercial fuel, the opportunity cost of which would be several billion USD.

(d) Hydrocarbon fuel (oil and natural gas) gradually replaced coal and lignite in total primary energy supply.

(e) Indigenous oil production could only meet 26% requirement.

(f) Production of coal in 2006–2007 touched 430.54 million tons.

(g) Distributional losses were more than 40%.

(h) Meeting energy requirement became complex in view of climate threats.

(i) Installed capacity of nuclear energy was only 3900 megawatt (3.1% of total installed power generation).

As expected, coal continued to dominate India's energy needs (50% share) with 78% of total domestic production of coal going into power generation. Globally, many nations continued to produce, use, import/export coal with impunity ignoring the warnings of scientific community that coal was responsible for global warming in a big way. China was the largest producer of coal (2226 million tons) followed by USA (951 million tons) and India (398 million tons). Several other countries like Australia (231 million tons), Indonesia (108 million tons), Russia (76 million tons), and South Africa (73 million tons) were major exporters, while others like Japan (178 million tons), South Korea (77 million tons), Chinese Taipei (61 million tons), UK (44 million tons), and Germany (38 million tons) were major importers (Planning Commission 2008).

From a long-term perspective of the growing threat of climate change and keeping in mind the need to diversify energy sources, renewables remained important to India's energy sector, especially rural India where 86% households continued to use firewood at the turn of the century. Only 5% of the households in rural areas and 44% households in urban areas used LPG; kerosene was used by 22% of urban households and only 2.7% of rural households for cooking. Firewood and dung cake dominated the household energy consumption in rural areas as can be seen in the following (Fig. 2.33).

It is surprising that no pan India systematic studies were carried out for more than six decades to assess the use of non-commercial energy, its environmental impact, direct and opportunity cost, and future potential. This form of energy continued its dominance in rural households for meeting their cooking and heating needs (Fig. 2.34). The consumption of 147.56 Mtoe of traditional fuels in 2006–2007 included consumption of 238 million tons of fuelwood, 98 million tons of dung, and 38 million tons of agricultural waste (Source—Planning Commission 2008). Around

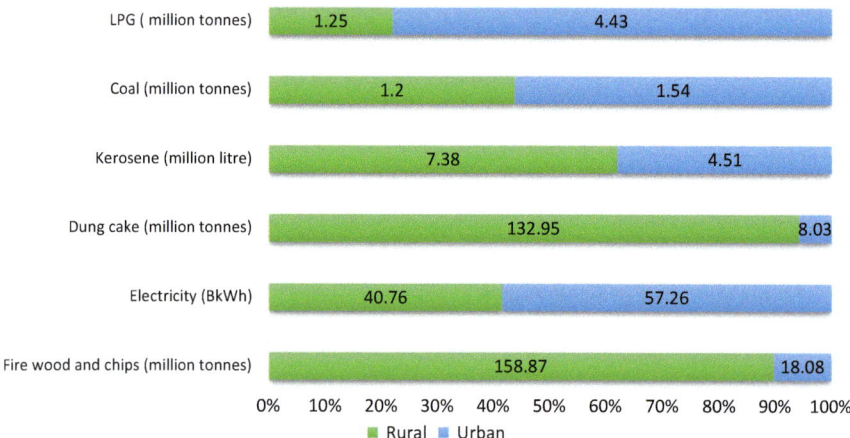

Fig. 2.33 Household energy consumption in India (July 1999–June 2000) (*Source* NSS 2001)

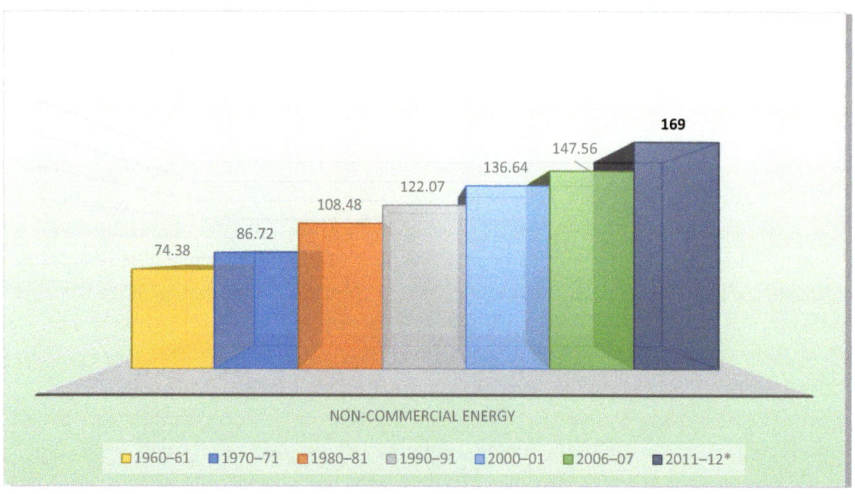

Fig. 2.34 Decadal growth of non-commercial energy (Mtoe) (*Source* Planning Commission 2008)

80% of the fuelwood was used for household consumption and the balance was used by the cottage industry and hotels. The noticeable feature of non-commercial energy use was its decline vis-à-vis total energy consumption (Fig. 2.35). Though the decline from 63.47% in 1960–1961 to 23.64% in 2011–2012 appears impressive for statistical record, but the fact that actual use continued its upward trend is perturbing (Source—Planning Commission 2008). This also indicates great rural–urban energy divide that has deterred India in fulfilling its climate-related commitments on the one hand and continue to impact environmental and human health on the other. The inefficient methods of using biomass-based fuels have not only resulted in low thermal

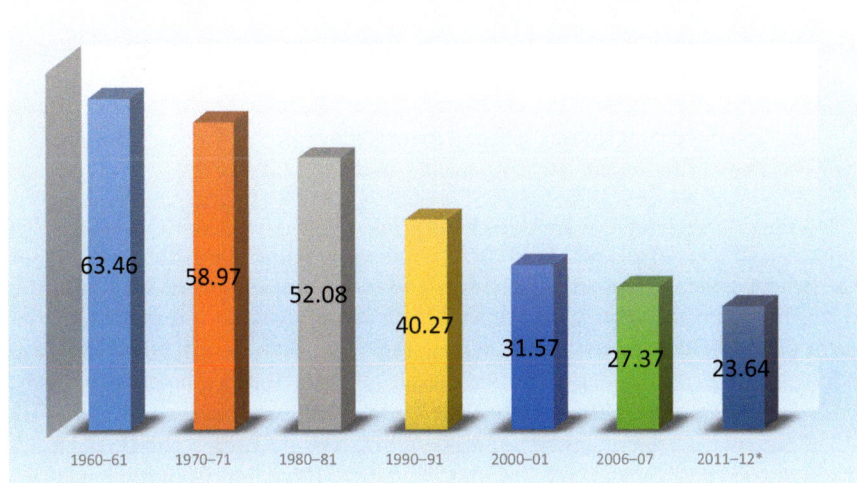

Fig. 2.35 Proportion of non commercial energy to total primary energy (in %) (*Source* Planning Commission 2008)

efficiency but high emission of air pollutants such as total suspended particulates, carbon monoxide, and formaldehyde as well. This has been and will continue to take heavy toll on the health of women and children with ailments such as bronchial asthma, bronchitis, TB, and chest infection. Risk from all respiratory diseases and eye diseases will undoubtedly increase with the length of use of traditional fuels. Use of animal and plant waste as fuel deprives the already deficient soil of its minerals and organic content. The collection of traditional fuels results in drudgery of women and children since they had to walk long distances to collect such fuels. Moreover, the time spent for this purpose is significantly long, depriving them of their additional income which they would have earned otherwise.

2.5.3 The Twelfth Five-Year Plan (2012–2017)

2.5.3.1 Highlights

(a) In 2016–2017, the share of coal and crude oil accounted for 64.17 and 31.25%, respectively.
(b) The industrial sector used 57.71% of the total energy consumption in 2016–2017.
(c) Import dependence of coal was estimated to be 22.4% and that of crude oil 78% by the end of 12th plan.
(d) Two-thirds of all households were using LPG as cooking fuel in 2009–2010.

(e) Per capita consumption of LPG in rural areas was 0.3 kilogram per month as compared to 1.8 kilogram in urban areas.
(f) Energy supply from renewable sources 24,503 megawatts by the end of the 11th plan.
(g) Total installed capacity as on March 31, 2012, including renewable energy sources of the country was 199,877 megawatts.
(h) The share of renewable energy capacity is about 12.2%.

For reasons elucidated in previous paragraphs, India remained heavily dependent on fossil and traditional fuels to propel its economy unlike many European nations that switched over to renewable resources and relied more on robust manufacturing sector. India, on the other hand, depended on its service sector as manufacturing sector's contribution to the GDP stagnated at 16%, thus raising questions about India's development strategy. India's energy intensity (energy input required for producing one unit of GDP) of 0.191 Kgoe/USD in 2011 was much above that of UK (0.102 Kgoe/USD), Germany (0.121 Kgoe/USD), Japan (0.125 Kgoe/USD), Brazil (0.134 Kgoe/USD), and USA (0.173 Kgoe/USD) (Planning Commission 2015). Unprecedented population growth, inadequate indigenous resources and funds, cost overrun, bilateral and international conflicts, archaic technology, and several other factors were cumulatively responsible for the state of affairs in energy sector.

What eventually matters in global competition is a high GDP and most of the developed nations ensured high contribution from manufacturing as well as service sector. Higher levels of GDP obviously require more energy, better efficiency, and round-the-clock availability. A comparison of GDP and population of several countries (Fig. 2.36) indicates in order to be economically strong, a nation must ensure higher levels of efficiency and technology use.

For India, the availability of electricity supply continues to remain an area of concern, particularly in rural areas, where consumers get supplies for less than 8 hours a day in certain states. Though 67% of the rural households are reported to have got access to electricity in 2009–2010, their per capita consumption of electricity was only around 8 units per month, which was one-third of reported consumption of 24 units in urban areas. This was due to erratic electricity supplies and reflects significant unmet demand. At the same time, access to LPG supplies in rural areas increased from 8.6% in 2004–2005 to around 15.5% in 2009–2010 but per capita consumption was only 0.3 kilogram per month as compared to 1.8 kilogram in urban areas. This disparity was due to partial dependence and use of LPG for various reasons, including cost of LPG, timely availability, and continued use of traditional fuels. Complete switchover to LPG may take several years of confidence building, sustained supply of LPG, mindset change, income enhancement, and adequate savings (Planning Commission 2015).

Not surprisingly, coal continued to be the dominant source of primary energy. Domestic production of coal and lignite accounted for two-thirds of total production of commercial energy in 2000–2001 and was projected to be about the same in 2021–2022. As a percentage of total consumption of commercial energy, the share of coal and lignite was projected to increase to 57%, from a level of 50% in 2000–2001. While share of oil in total commercial energy consumption was expected to decline from 37.5% in 2000–2001 to 23.3% in 2021–2022, the share of natural gas and

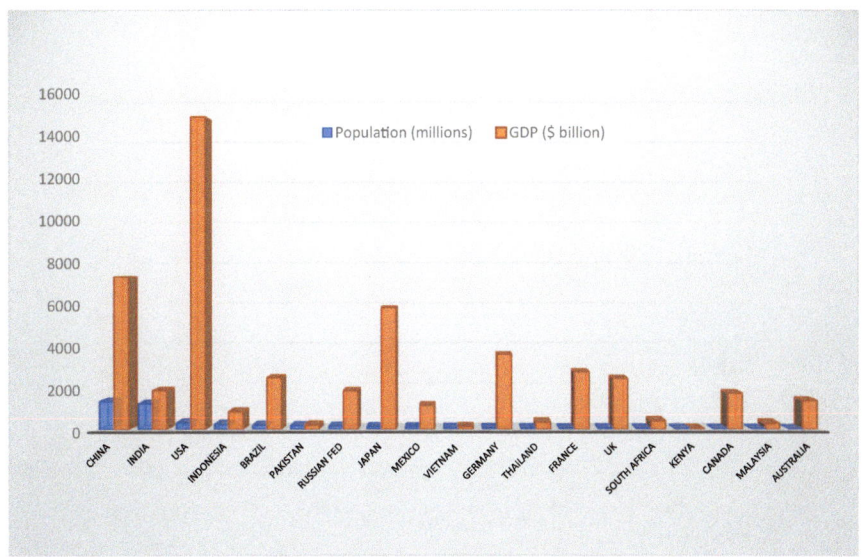

Fig. 2.36 A comparison of country population (million) and GDP (USD Billion) (*Source* World Bank 2013)

liquefied natural gas (LNG) was anticipated to rise from 8.5% to 13% during the same period. The combined share of oil and natural gas in energy consumption was 24.7% in 2011–2012 and was expected to be about the same in 2021–2022 (Planning Commission 2015).

Figure 2.37 shows the trend in the use of different fuels for two decades starting 2000–2001. Non-commercial fuels and coal dominate the trend with oil and natural gas playing supportive role. Hydropower has its limitations. Normally it takes long time to operationalize a hydel plant, and the normal life expectancy is about 30–35 years after which they need modernization. Many existing hydropower stations require modernization to generate reliable and higher yield.

The question of sustainability of thermal power stations based on present-day subcritical technology (with an efficiency of about 38%) had to be addressed to improve energy efficiency and reduce the emission of GHG and other obnoxious gases. It was decided that all new thermal power plants would be based on supercritical technology. Eleven supercritical units with a total capacity of 7,400 megawatts are already functional and a number of supercritical units were under construction.

The 12th plan also recognized several deficiencies in village electrification program. Nearly 6,000 villages electrified till December 2011 could not be energized due to lack of supporting network or other resources. The plan admitted that access to electricity in rural areas especially small hamlets was inadequate, poor financial health of utilities and high cost of power were disincentive for new connections, absence of supporting network in some state and a viable revenue model was wanting.

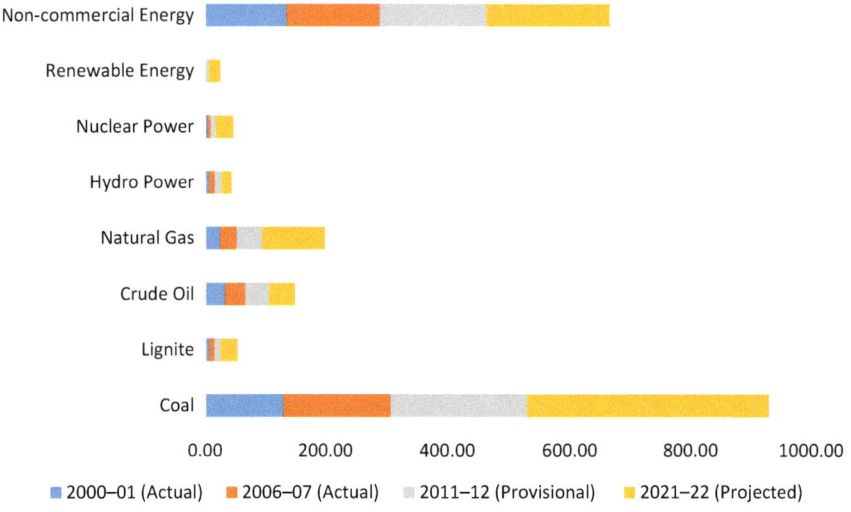

■ 2000–01 (Actual)   ■ 2006–07 (Actual)   ■ 2011–12 (Provisional)   ■ 2021–22 (Projected)

Fig. 2.37 Trends in supply of commercial and non commercial energy (in mtoe) (Planning Commission 2015)

Other areas of concern as mentioned in the 12th plan included:

i. In certain states, even the minimum required hours of supply (of 6–8 hours) could not be met.

ii. Though 67% rural households had access to electricity in 2009–2010, their per capita consumption was only around 8 units per month which was one-third of consumption of 24 units in urban areas. This is because of poor quality of electricity supplies and reflects significant unmet demand.

iii. Access to LPG supplies in rural areas increased from 8.6% in 2004–2005 to around 15.5% in the year 2009–2010. Nonetheless, per capita consumption in rural areas was just 0.3 kilogram per month as compared to 1.8 kilogram in urban areas.

iv. There were frequent complaints of burning of transformers, which is an indication for upgradation of transformer capacity as the current average demand of below poverty line (BPL) and above poverty line (APL) consumers was in the range of 300–500 watts and 0.5–1.15 kilowatt, respectively.

v. Connections to APL households were slow on account of poor supply of electricity, long delays in processing of applications, and inadequate transformer capacity.

vi. In most of the operating states, no franchisee was found in any of the surveyed villages and the distribution companies had their own mechanism of meter reading and billing.

vii. Electrification failed to generate substantial employment opportunities or economic development in the rural areas except in a few cases.

viii. Clean development mechanism for utilization of fly ash, control of sulfur and nitrogen oxides and mercury in coal-based thermal power plants required immediate attention for clean and green energy.

ix. High cost of renewable energy (Fig. 2.38) especially solar was a disincentive for rural areas despite huge potential of generation:

Power generation from renewables was projected to increase to 54,503 megawatts by the end of 12th plan and 99,617 megawatts by the end of 13th plan. But the share of renewables in total commercial energy use was expected to rise from about 1% in 2011–2012 to 1.43% in 2016–2017 and just under 2% in 2021–2022. Considering the fact that soon India will overtake China in population, there is an urgent need to improve renewable energy potential vis-à-vis Brazil, USA, Canada, and China (Fig. 2.39).

Source	Estimated initial capital cost (INR in crore/MW)	Estimated cost of electricity generation (INR/kWh)
Small Hydro Power	5.50 – 7.70	3.54 – 4.88
Wind Power	5.75	3.73 – 5.96
Biomass Power	4.0 – 4.45	5.12 – 5.83
Solar Power	10.00 – 13.00	10.39 – 12.46
Reference – [Plan. Comm., 2015]		

Fig. 2.38 Financial cost of renewable energy

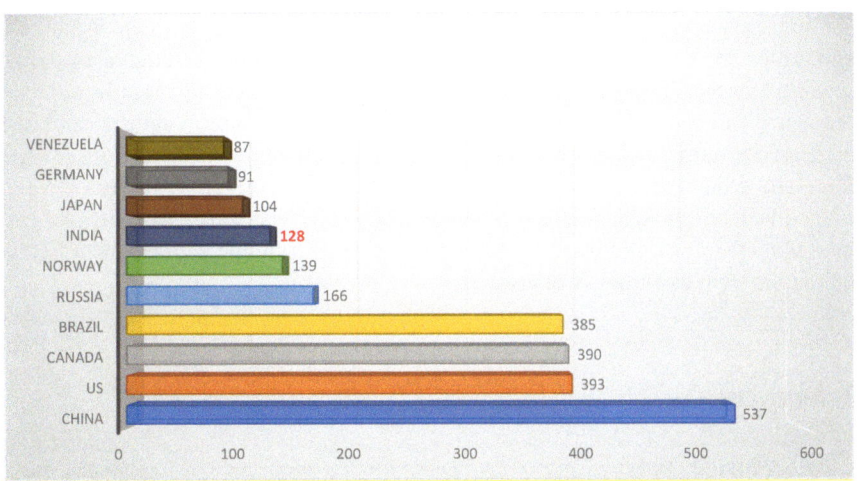

Fig. 2.39 Electricity production from renewable resources (billion kilowatt hour) (Folger, T. "National Geographic Magazine." (2011))

The 12th plan made the following suggestions for energy security:

i.   Enhancement of domestic production of coal, oil and gas, and other energy sources was urgently required and for this purpose availability of land, clearances for environment and forest, and implementation of the Scheduled Tribes and Other Traditional Forest Dwellers (Recognition of Forest Rights) Act, 2006 should be expedited;

ii. Uncertainty about production sharing contracts should be appropriately addressed.

iii. Devise management strategies and procedures for ensuring effective implementation of fuel development projects.

iv. A viable and attractive policy regime for substantial private investment including foreign investment in oil and natural gas blocks and new capacities for renewable energy.

v.   Investments in renewable energies must be enhanced as the rate of share of renewable energy in total energy consumption will only reach 2% by 2021.

vi. Investments in energy assets in foreign countries, especially for coal, oil, and gas and uranium should be stepped up.

vii. As a shield against disruption in oil supplies (India is import-dependent to the extent of more than 80%) storage capacities need to be created. Increasing storage capacity has not been the priority of Indian government and only 5.33 million tons of petroleum products against annual consumption of 183.5 million tons in 2015–2016 was available. The Organization for Economic Cooperation and Development (OECD) countries have generally created these capacities to the extent of 90 days of their domestic demand.

By 2040, almost a century after independence, India endeavors to attain energy consumption almost equivalent to those of the developed countries. In order to realize this aim, new and emerging technologies have to be adopted in order to harness renewables such as solar, wind, and bio. Next two decades are crucial to define a realistic energy pathway as climate-induced disaster events increase in number, frequency and intensity. There will be international pressure to reduce coal and crude oil consumption and switch over to natural gas, solar, wind, and hydro energy. There are strong indications that transport sector will be driven by lithium ion and solar cells. Household and industrial sector will be dominated by electricity and gas. India may be left behind by miles should it decide to continue with coal and lignite in the long term as currently envisaged.

References

Aayog NITI (2017) Draft national energy policy. National Institution for Transforming India, Government of India, New Delhi. https://niti.gov.in/writereaddata/files/new_initiatives/NEP-ID_27.06.2017.pdf

Bhattacharyya SC (2005) The Electricity Act 2003: will it transform the Indian power sector? Utilities Policy 13(3):260–272

Central Electricity Authority, Ministry of Power, Government of India (2018). https://www.cea.nic.in/

Dandekar VM, Rath N (1971) Poverty in India-I: dimensions and trends. Economic and Political weekly 25–48

Energy Statistics (2018) Central Statistics Office, National Statistical Organization, Ministry of Statistics and Programme Implementation, Government of India

Lok Sabha (2019) Lok Sabha question 337 (Archive)

Mahalanobis PC (1955) The approach of operational research to planning in India. Sankhyā: The Indian Journal of Statistics (1933–1960) 16(1/2):3–130

Ministry of Environment and Forests (MOEF) (1999) National forestry action programme—India, vol 1: Status of Forestry in India 79

NSS (2001) 55th round. National Sample Survey Organisation Ministry of Statistics & Programme Implementation Government of India, New Delhi

Planning Commission (1953) First five year plan

Planning Commission (1956) Second five year plan. Planning Commission. India

Planning Commission (1961) Third five year plan. Summary

Planning Commission (1970) The fourth five year plan 1970–75

Planning Commission (1976) Fifth five year plan 1974–79

Planning Commission (1979) Report of the working group on energy policy. Planning Commission, Government of India, New Delhi

Planning Commission (1980) Sixth five year plan 1980–85

Planning commission (1985) The seventh five year plan, 1985–90. Government of India Press, India.

Planning Commission, Government O. F. India (1992) Eighth five year plan (1992–1997) V. II. Government of India, New Delhi

Planning Commission (1997) Ninth five year plan 1997–2002

Planning Commission (2001) Indian planning experience-A statistical profile

Planning Commission (2002a) Report of the committee on India Vision 2020

Planning Commission (2002b) Tenth five year plan 2002–07

Planning Commission (2008) Eleventh five year plan 2007–12

Planning Commission (2015) 12th five year plan (2012–17)

Thakur T, Deshmukh SG, Kaushik SC, Kulshrestha M (2005) Impact assessment of the Electricity Act 2003 on the Indian power sector. Energy Policy 33(9):1187–1198.

UNICEF, WWF (1998) Freshwater for India's children and nature. New Delhi.

World Bank (2013) Little green data book 2013. World Bank

World Bank (2017) Little green data book 2017. World Bank

Chapter 3
The Challenges of Energy Supply

Abstract Energy, both internal and external, is required at every stage of human life. The formative years count on food and oxygen for survival and growth, whereas the latter part of life is more dependent on external energy sources. Experience so far indicates that both internal as well as external energy play crucial part in human development. Since independence, India's record of health, education, and access to commercial energy has been unsatisfactory and it's time to adopt out of the box strategy for ensuring high energy supply to achieve and sustain higher trajectory economy. There are many impediments for improving the energy productivity including raw material (for nuclear energy), technical feasibility, finances, public opposition and demonstrations, water and land rights, environmental clearances, and legal cases for thermal and hydro that need to be resolved to exploit full potential.

Keywords Per capita electricity consumption · GDP and energy · Household electrification · Electrified villages · Energy access and security

3.1 The Understanding and Perception

The word 'energy' referred to in this chapter encompasses all kinds of energy tapped from various sources for the survival, growth, and development of human beings and is not restricted to electricity generated and supplied by government, individuals, and groups, either public or private sector.

All living beings on this planet need energy from the beginning of their creation. We start breathing as soon as we are born, and the oxygen so absorbed by iron molecule (of hemoglobin) starts energizing the body. In plants the role of iron is substituted by magnesium which forms the core of chlorophyll. In layman's understanding, the process of human development is associated with the health of a living body, that is, the quality and quantity of food an individual takes that helps in physical and mental growth and prevents malfunctioning of the body as well as premature death. Medical science tells us that if we can overcome any health issues during the initial years of our lives, the chances are that we will have longer life and better resistance in the later years.

© Springer Nature Singapore Pte Ltd. 2021
A. Srivastav, *Energy Dynamics and Climate Mitigation*,
Advances in Geographical and Environmental Sciences,
https://doi.org/10.1007/978-981-15-8940-9_3

For the sake of understanding, let us divide our life into three phases:

- Phase 1 (Health Phase): One must stay healthy so long as one is alive and to ensure good health, we must be healthy from the day we arrive on this planet. Human development indicators established by international organizations attach lot of importance to health of individual, especially in the early phase of life. In India we maintain data of neonatal and under five mortality. Despite large number of health schemes of the government, individuals especially in rural and peri-urban areas are still not conscious of health and hygiene, quality of food and water. Many poor children do not have access to regular breakfast. They go to school either empty stomach or at best after consuming a cup of tea. Besides tea, consumption of health retarders such as tobacco, bidi, cigarettes, local brewed and adulterated and poor-quality alcohol is widespread in India. The consequences are well known in medical fraternity and documented in research reports.
- Phase 2 (Education/Learning/Skill Phase): Education in India usually starts at the age of 4 years and the legal age for entry into class 1 is 6 years. From then on, an individual can pursue studies and gain knowledge as well as skills till about 25 years of age. These formative years are crucial for mental as well as physical development and requires formidable quantity of energy inputs.
- Phase 3 (Output Phase—Production of Service/Goods): This phase generally lasts for 40–50 years when individuals contribute to the GDP or GNP. This is the phase where large amount of energy is generated as well as consumed either individually or collectively. Inputs received during phases 1 and 2 collectively support this phase. Failure of a nation to provide adequate quantity of energy (external and external) to its population during output phase may have serious economic and social consequences. In Indian context it may be seen that from first five-year plan (1951–1956) the government laid emphasis on energizing tube wells and providing water for irrigation to ensure adequate supply of food. Out of the total first plan outlay of Rs. 2068 crores, almost 50% were spent on energy sector (Rs. 561.41 crores for irrigation and power and Rs. 497 crores for transport and communication). Many initiatives have been taken by the federal government as well as state governments from time to time, including green revolution, mid-day meal, food for work, rural employment guarantee, right to education, and skill development programs to enhance economic growth.

In the same way as food, energy in the form of electricity is also recognized as a significant input improving living standards, and there is a strong correlation between per capita electricity consumption and Human Development Index (HDI). Ironically, of the 1300 million Indians, almost a quarter (304 million) are without access to electricity and more than 38% (500 million) depend on non-commercial biomass-based fuel for cooking (NITI Aayog 2017).

The Little Green Data Book of the World Bank (2017) indicates that the per capita electricity consumption in India as 805 kilowatt-hour/year. In other words, an average Indian gets 2.2–2.8 units of electricity per day. These are notional figures and there are huge variations in electricity consumption across different states of India. Bihar, Nagaland, Tripura, Assam, and Manipur are the worst sufferers with

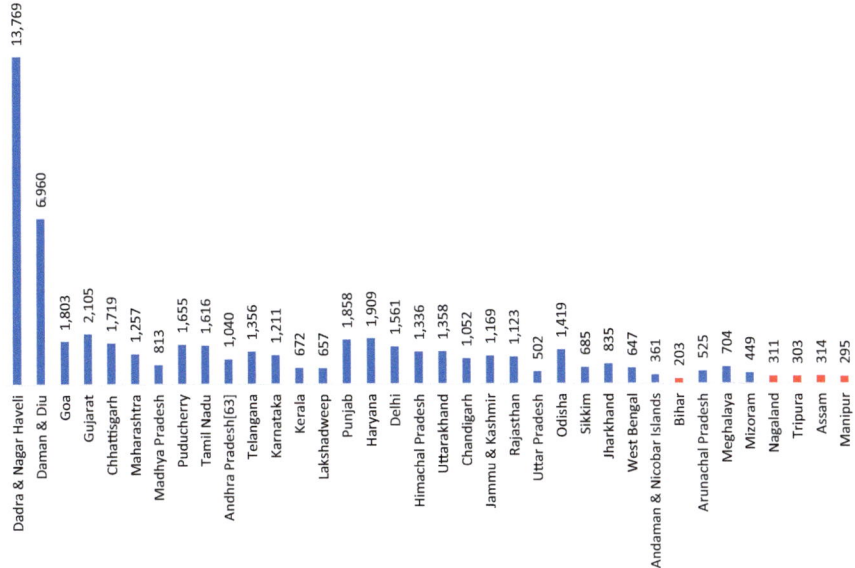

Fig. 3.1 Per capita electricity consumption (kWh) in different states of India in 2014–2015 (Rajya Sabha 2016)

per capita annual consumption ranging between 203 and 314 units only (Fig. 3.1). Compare this with global per capita electricity consumption of 3144 units or with 707 units in South Asia or 703 units of Zambia or with 217 units of Yemen or 140 units of Nepal or with Sri Lanka with per capita annual electricity consumption of 531 units or with Pakistan with per capita electricity consumption of 472 units or with USA with per capita electricity consumption of 12,973 units (World Bank 2017). The list is endless. India must decide its short, medium, and long-term goals of per capita consumption in a world of superfast mechanization where no nation with poor electricity or energy consumption can ensure and sustain higher trajectory economy. Achieving electricity connection in every village has little meaning unless the energy so supplied gets translated into real GDP through the process of education, health, technology, and skill enhancement. One good example of this can be seen in many villages of Gujarat where people have set up diamond cutting units in villages[1] and obtain small diamonds through sub-contract.

While installed electricity generation capacity (Fig. 3.2) in India, especially from thermal and renewable sectors, has shown quantum jump during the decade of 2008–2018, but the same has not been translated proportionately to household sector. Village household sector is way behind in per capita consumption for a variety of reasons. A large portion of India's poor population lives in villages and they have neither capacity nor willingness to pay electricity charges. Such households will

---

[1]Cutting of diamond is a high skill job that requires specialized knowledge, tools, equipment, and techniques.

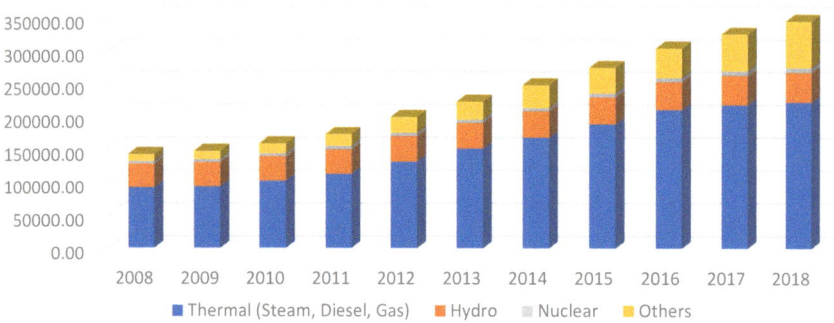

Fig. 3.2 Installed electricity generation capacity (MW) (Energy Statistics 2018)

be more than happy to use electricity free of cost. But then freebies are always unsustainable. Experience has shown that urban households used to adopt range of malpractices for reducing/not paying their electricity bills. In every likelihood, this practice will be adopted by rural households as well till such time they have enough savings. Moreover, disproportional rise in population is a serious impediment in managing household connectivity and recovery of use cost since millions of poor families are mobile (for jobs) for a large part of year.

It may not be appropriate to draw a conclusive relationship between population, GDP, and energy supply, yet a comparison of data from several countries (World Bank 2013) indicates the following:

1. Countries using high proportion of fossil fuel for electricity generation must aim at improving conversion ratio of GDP to fossil fuel (Fig. 3.3).
2. Higher GDP/population ratio indicates efficiency, technology use, and higher per capita income (Fig. 3.4).
3. For countries where use of fossil fuel is disproportionately high (Fig. 3.5) and will remain so for a few more decades, it is imperative to reduce population at a faster rate, introduce new technology, and improve energy efficiency to scale up GDP. It is equally significant to understand that the average life of a thermal power plant is 40 years and many, if not most, of India's coal-based thermal power plants are several decades old and need to be overhauled at the earliest. Moreover, the gross calorific value of Indian coal used for power generation is 3541 kcal/kg whereas that of imported coal is 5500 kcal/kilogram. Outages are common and the same is evident from the fact that in September 2018 and October 2018, 26 and 22 thermal power stations, respectively, suffered outages due to shortage of coal supply (Lok Sabha 2018).
4. Countries must overhaul their existing sub-critical thermal plants to super and ultra-super as the old plants are living hell and spew venomous gases as shown in Fig. 3.6.

However, the same plant releases the following gases and other wastes during the process of electricity generation (Fig. 3.7).

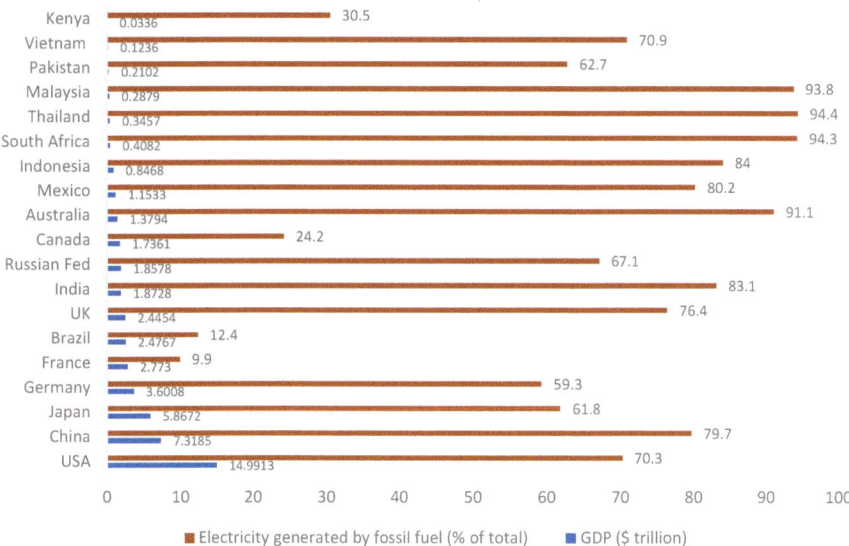

Fig. 3.3 Comparison of GDP (trillion USD) and use of fossil fuel in electricity generation (%) (*Source* World Bank 2013)

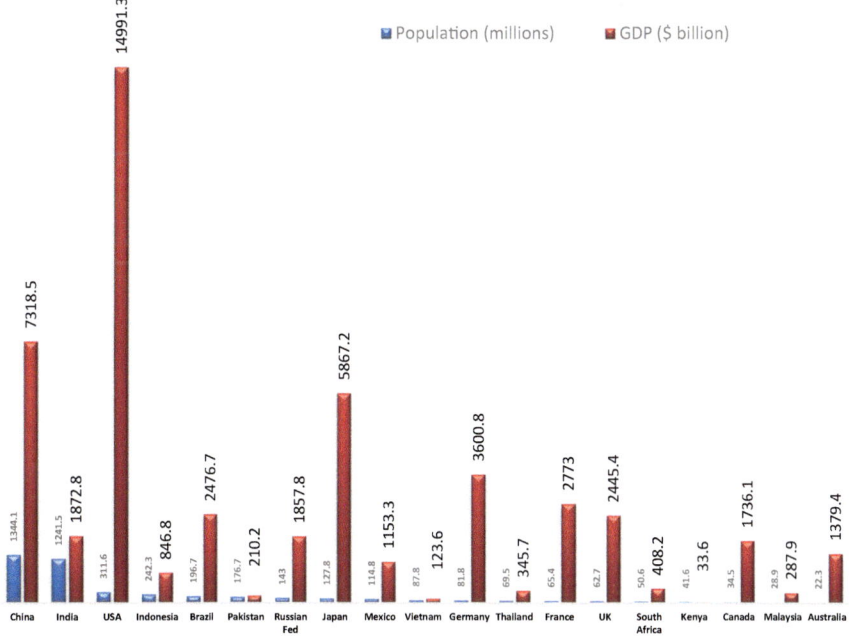

Fig. 3.4 Comparison of country population (million) and GDP ($ billion) (*Source* World Bank 2013)

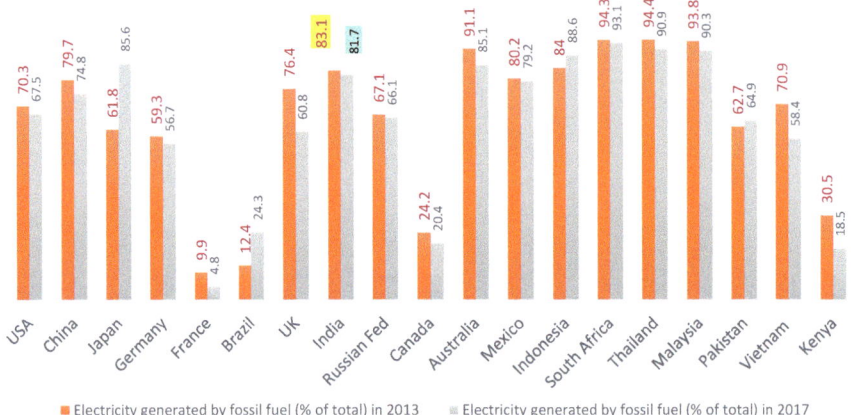

■ Electricity generated by fossil fuel (% of total) in 2013   ■ Electricity generated by fossil fuel (% of total) in 2017

Fig. 3.5 Electricity generated by fossil fuel (%) (*Source* World Bank 2013, 2017)

Fig. 3.6 A 500-MW thermal plant uses limestone, coal, and water (*Source* Srivastav 2019)

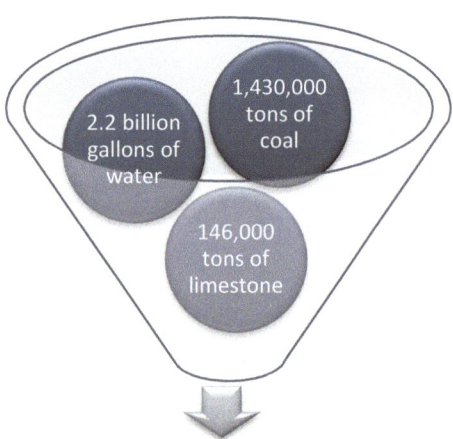

To Produce 3.5 billion kilowatt-hours per year.

3.2 Carbon-Intensive Energy Sector

India's energy sector accounts for 77% of its greenhouse gas emission and 4.91% of global greenhouse gas emission (Sinha et al. 2019). Heavy dependence on coal for power generation and the use of inefficient sub-critical plants to burn it push up the carbon intensity of India's power sector to 791 g of carbon dioxide per kilowatt-hour (gCO_2/kWh), compared to the world average of 522 gCO_2/kilowatt-hour. Based on past emission levels and GDP forecast, absolute emissions may go as high as 6.5 billion tons of carbon dioxide emissions in 2030 (Sinha et al. 2019) as compared with 3.5 billion tons in 2015, a rise of almost 85%.

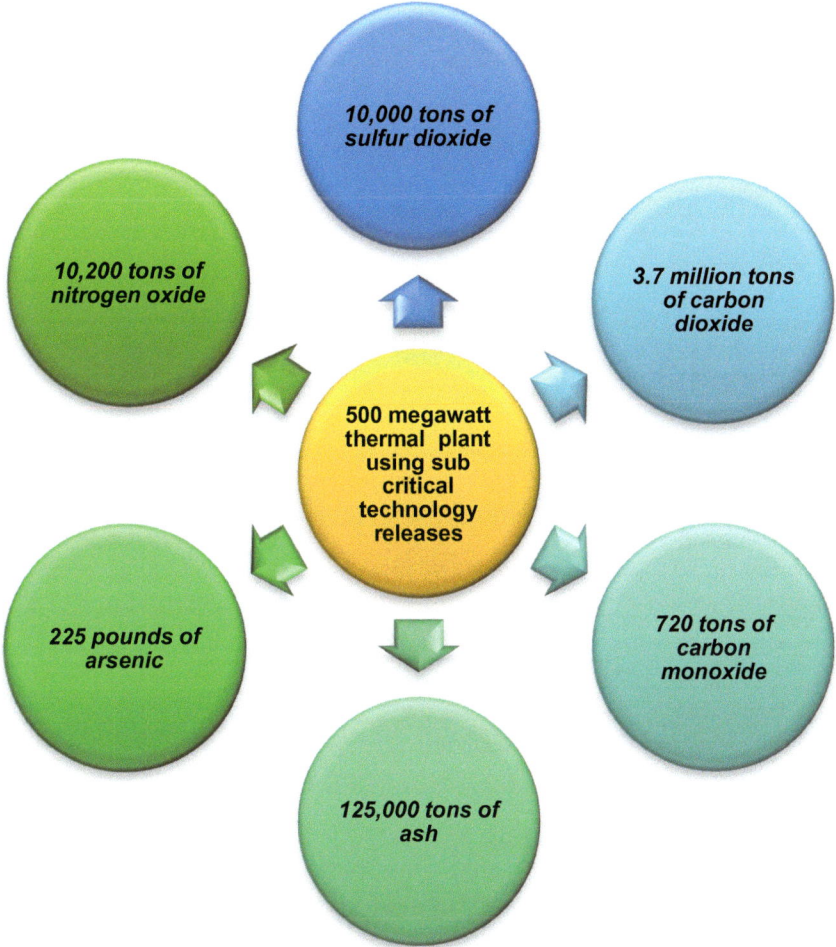

Fig. 3.7 Waste generated by thermal power plant (Srivastav 2019)

Being a frontrunner in global economy, India must address the challenges that energy sector faces (and not confine to providing electricity alone), especially in view of the climate-related risks. One major advantage this country has over others is the fact that the basic infrastructure like smart cities, energy grids, buildings, and roads are in the process of being set up, and hence it will be easier to switch over to low-carbon/zero-carbon trajectory. Inclusion of natural gas in energy basket in recent years has helped to reduce the carbon intensity of electric power generation. Besides, improved combustion efficiency of conventional coal technologies and strong promotion of renewable technologies have also made appreciable progress.

Fig. 3.8 Capacity (MW) and actual generation (million units) of electricity in India during 2018–2019 (till December 2018) (*Source* Lok Sabha 2019a)

The 12th five-year plan (2012–2017 period) is the first plan that took climate risk seriously and suggested the following focal areas (Planning Commission 2015), namely:

i.   Advanced coal technologies.
ii.   National wind and solar energy mission.
iii. Technology improvement in steel and cement industry.
iv.   Energy efficiency program in industry.
v.   Vehicle fuel efficiency.
vi.   Efficiency in freight transport.
vii. Faster adoption of green building codes.
viii. Improving the stock of forest and tree cover.
ix.   Enhance 30,000 megawatts of renewable energy capacity (Fig. 3.8).

3.3 Inconsistent Power Generation by States

During the decade of 2006–2016, several states/union territories showed decline in electricity generation, including national capital Delhi, where electricity generation declined from 10,979.6 MWh in 2006 to 6206.1 MWh in 2016. Inexplicably, the state of Goa showed drastic reduction from 443 MWh in 2006 to zero MWh in 2016. Similarly, Dadra and Nagar Haveli, Chandigarh, Lakshadweep, and Mizoram brought down their generation to naught.

The details are provided in Fig. 3.9.

Fortunately, the downtrend in electricity generation by some states was compensated by sharp increase in production in 20 states/union territories with Sikkim showing a jump of 820% from 386 MWh in 2006 to 3551.9 in 2016, followed by Tripura 355% and others as shown in Fig. 3.10.

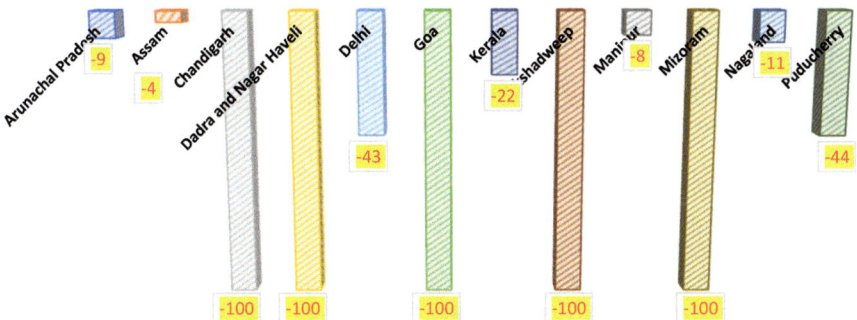

Fig. 3.9 Decline (%) in electricity generation (MWh) between 2006 and 2016 (*Source* NITI Aayog 2015)

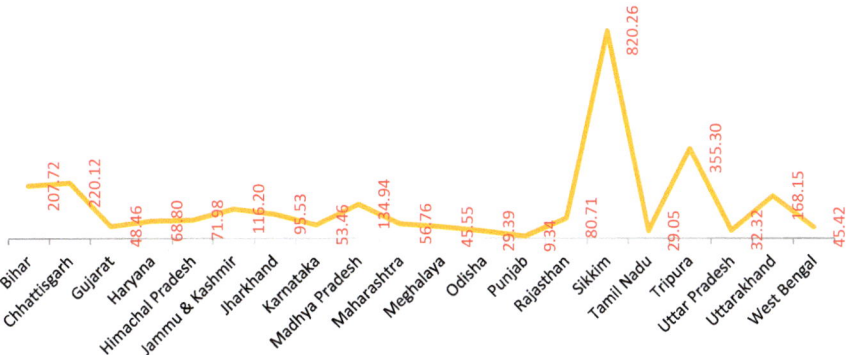

Fig. 3.10 Increase (%) in electricity generation (MWh) between 2006 and 2016 2016 (*Source* NITI Aayog 2015)

Attempts have been made in recent years to improve the supply challenges and this has resulted in reducing the energy shortages. Figure 3.11 shows that the increasing trend in energy shortage that had reached 86,905 million units in 2012–2013 (largest

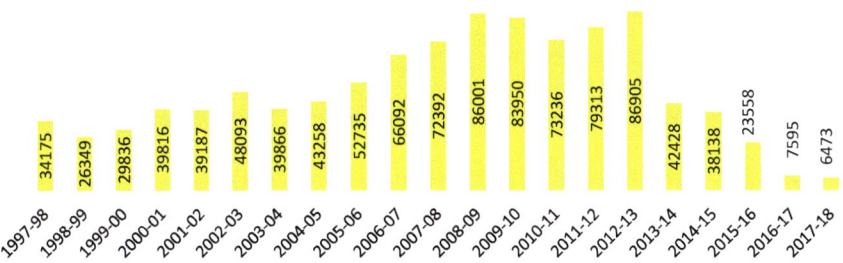

Fig. 3.11 Energy shortage (million units) (*Source* Ministry of Power 2018)

since 1997–1998) has been arrested with extraordinary performance after 2015–2016.

While this may sound satisfactory, the Energy Performance Index (Figs. 3.12 and 3.13) developed by the World Economic Forum to assess the performance of energy systems of individual countries using three indices, viz., economic growth and development, environmental sustainability, and energy access and security ranks India at 87th position among 127 countries. India is no better than Jamaica and South Africa and the fact that countries like Vietnam, Armenia, and Peru have better ranking should be a matter of concern. Unless we bring in energy revolution and improve our performance in coming decades by reducing carbon dioxide emissions, vehicular pollution, agricultural crop burning, refuse combustion and fireworks, and replace biomass-based fuels used for cooking, we are certain to slip back in ranking. India must strive hard for scoring more than 0.8 as soon as possible.

The following quote from EAPI report (WEF 2017) should serve as the benchmark for future energy strategy:

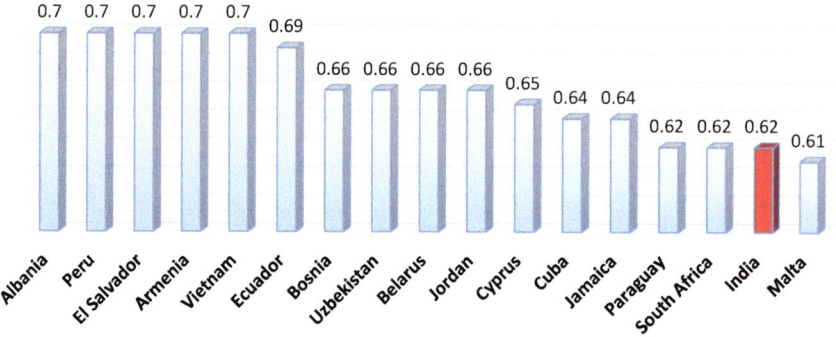

Fig. 3.12 Energy access and security score (2017) (*Source* WEF 2017)

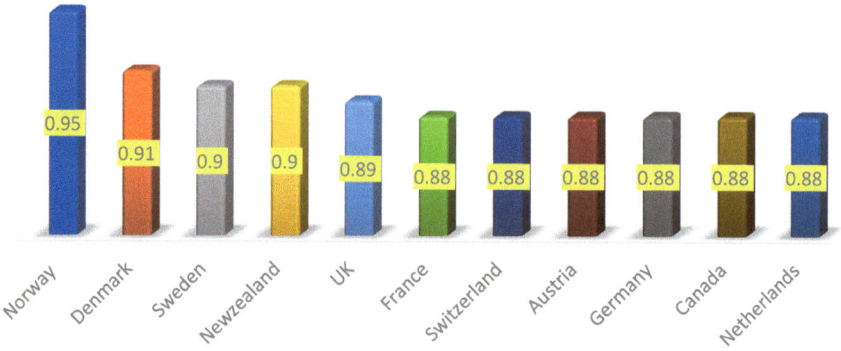

Fig. 3.13 Energy access and security score (2017) of developed countries (*Source* WEF 2017)

India (87th) is gradually improving its performance on the EAPI (90th last year). Similar to China, the country boasts a strong score on the indicator for diversification of import counterparts (5th), but its energy system continues to face some significant challenges, particularly in environmental sustainability (109th). India has some of the lowest scores in the EAPI for CO_2 emissions from electricity production and PM2.5 levels (117th and 123rd, respectively). While sources of pollution are diverse and intermittent (e.g. agricultural crop burning, refuse combustion, fireworks), the energy sector is a large, consistent contributor to this issue of major concern. Many solutions have been attempted with varying degrees of impact, but the country sorely needs a comprehensive plan of action to implement an effective and sustainable answer.

India also faces an uphill battle to increase energy access and security (95th). A large percentage of the population still lacks access to electricity (101st) and uses solid fuels for cooking (108th).

3.4 The Fallacies of Household and Rural Electrification

According to the census report of 2011 (Chandramouli 2011), there were 330,886,373 (33 plus crore[2]) houses in the country of which 92.5% were occupied. The term housing in census record incorporates residential buildings, shops, offices, hotels, guest houses, hospitals, dispensary, and so on. The breakup is provided in Fig. 3.14.

As per this record, there were 23.61 crores residential buildings in 2011. However, the details submitted (Fig. 3.15) by the government to the Lok Sabha (House of the People) in November 2018 (Lok Sabha 2018) mentioned that among 21.73 households in the country, 20.79 crores were electrified. Believing that no new residential building has been added after 2011, there will still be 30 million residential buildings without electricity connectivity (Fig. 3.16). There are many states which are still below 80% mark such as Assam, Bihar, Jharkhand, Manipur, Meghalaya, Nagaland, Odisha, and UP. In addition to the discrepancy in number of residential building (that must have increased by millions after 2011) and tardy progress of some states, there are issues of sustainability of electricity supply and replacement of biomass fuels.

Energy poverty has its cost (in terms of foregone development) that India has paid and is likely to pay in future as well. A total of nearly 304 million people still suffer from want of electricity and there is no reliable data on the number of households or people who do not have adequate electricity. Lack of access to electricity or inadequate electricity in homes impacts education, health, and economic development, especially in rural India. Like urban consumers, the rural households have equal right to get connection and receive efficient services. The national energy policy aims at achieving 100% electrification by 2022 and will take this to be the main plank of the overall energy strategy. The principle vehicle for achieving 100% electrification goal is through Deen Dayal Upadhyay Gram Jyoti Yojana (NITI Aayog 2017). Studies have revealed that despite major strides made through government programs, the problem of electricity access did not improve appreciably. An inherent challenge in the process is ensuring the coverage of households as opposed to only villages.

---

[2]One crore = ten million.

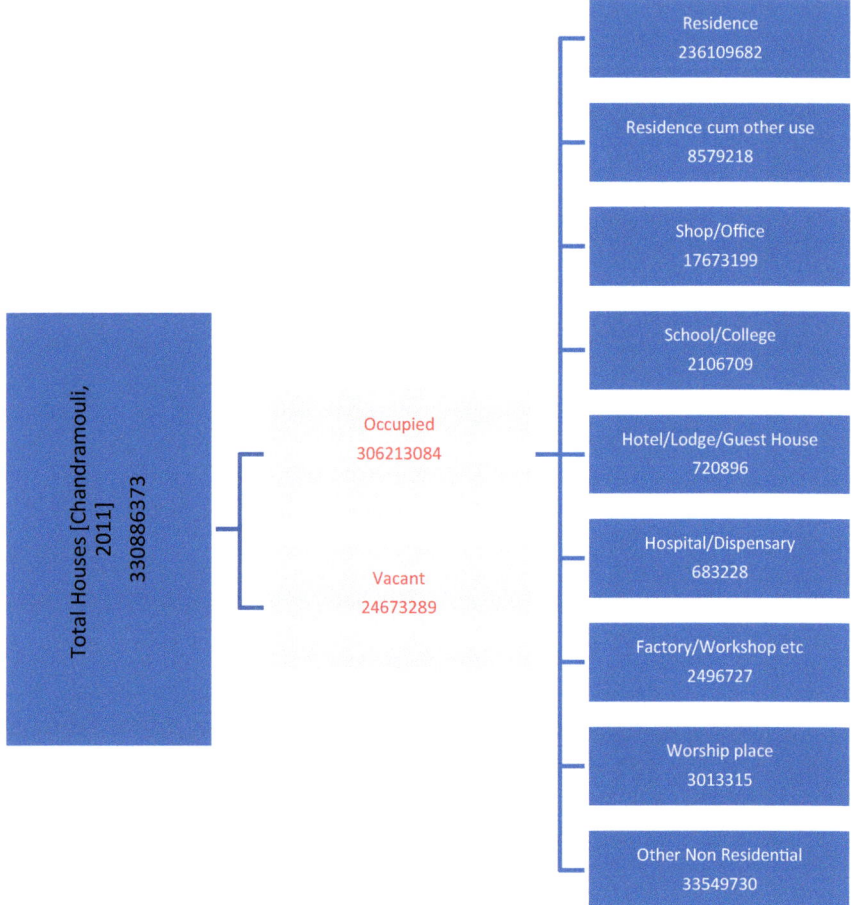

Fig. 3.14 Total Houses (Chandramouli 2011)

Several states with high electrification rates still have poor household electrification, and certain hamlets, not covered in the national sample surveys and the DDUGJY, housing a considerable section of the populace, are also without access to energy.

There is another dimension to household electrification. The current definition of an electrified village (Box 1) is flawed as it does not cover all households.[3] Government will have to redefine the concept of 'Electrified' village and a village should be

---

[3](Source—Ministry of Power—Letter No. 42/1/2001-D(RE) dated February 5, 2004 and its corrigendum vide letter no. 42/1/2001-D(RE) dated February 17, 2004). Website https://www.ddugjy.gov.in/page/definition_electrified_village).

As per the new definition, a village would be declared as electrified, if,

• Basic infrastructure such as distribution transformer and distribution lines are provided in the inhabited locality as well as the Dalit Basti hamlet where it exists;

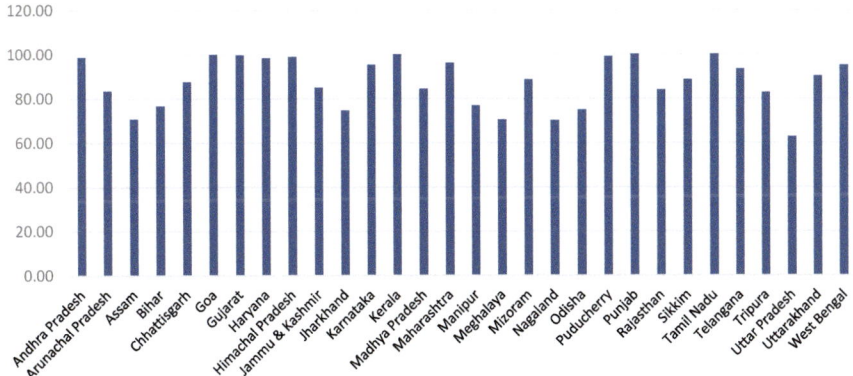

Fig. 3.15 State-wise achievement (%) in household electrification (Lok Sabha 2018)

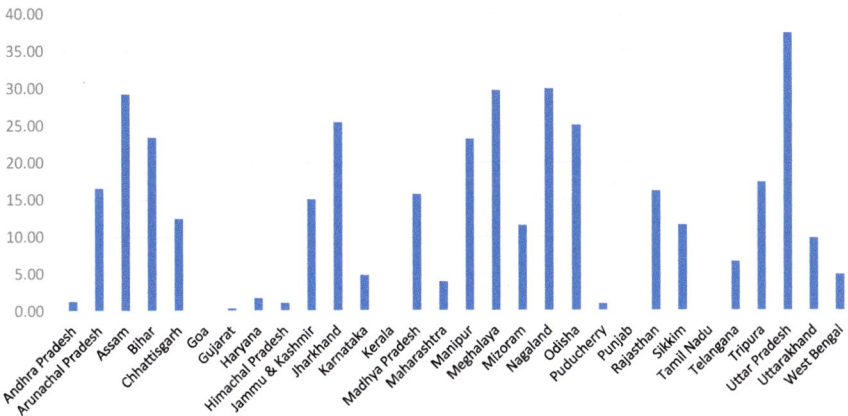

Fig. 3.16 Shortfall (%) in household electrification (Lok Sabha 2018)

deemed electrified only when all households of that village have electricity connection with reliable supply of electricity at least for a set number of hours. Although rural consumers may be extended capital subsidy for initial connectivity, as is being done through DDUGJY, the same should be equitable, well targeted, and have an exit strategy, with the consumer being reckoned as a revenue-generating variable thereafter. The non-BPL households may also be offered support, albeit at a different level. In areas where the reach of the grid is limited due to geographical constraints, incentivizing micro-grids makes economic sense. Micro-grids may play another role,

---

- Electricity is provided to public places like schools, panchayat office, health centers, dispensaries, community centers, etc.
- The number of households electrified should be at least 10% of the total number of households in the village.

even in electrified villages. They offer a value proposition in meeting peak electricity requirement in electrified villages.

Box 1 - Evolving concept of "Electrified Village" in India

Pre October 1997 period — A village was classified electrified if electricity was used within its revenue area for any purpose whatsoever.

Post October 1997 period — A village was classified electrified if electricity was used in the inhabited area within the revenue boundary of the village for any purpose whatsoever.

Post February 2004 — A village would be declared electrified if

> Basic infrastructure such as distribution transformers and lines are provided in the inhabited locality.
> Electricity is provided to public places such as schools, panchayat office, health centers, community centers, etc
> The number of households electrified should be at least 10% of the total number of houses in the village.

Concepts and definitions used in village and town directory as per District Census Handbook

Power Supply Categories:

1. For domestic use: electricity used of domestic consumption only.
2. For agricultural use: electricity connection given to farmers for farming and irrigation.
3. For commercial use: electricity connection given for industries and other commercial activities.
4. For all other use: electricity connection given for domestic, agricultural, and commercial purposes.

Connections to residential houses, bungalows, clubs, hospitals, and hotels run on noncommercial basis, charitable, education and religious institutions are included in domestic category.

[Source - Ministry of Power — Letter No. 42/1/2001-D(RE) dated 5th February 2004 and its corrigendum vide letter no. 42/1/2001-D(RE) dated 17th February 2004). Website .http://www.ddugjy.gov.in/page/definition_electrified_village]

There is another dimension to household electrification—urban slums.[4] These have mushroomed in all metropolitan cities with extremely high population density. There were 4041 statutory towns, 2543 slums, and 13,749,424 slum households

---

[4]What constitutes a slum:

1. All notified areas in a town or city notified as 'Slum' by state, UT administration, or local government under any Act including a 'Slum Act'.
2. All areas recognized as 'Slum' by state, UT administration, or local government, housing and slum boards, which may have not been formally notified as slum under any act.
3. A compact area of at least 300 population or about 60–70 households of poorly built congested tenements, in unhygienic environment usually with inadequate infrastructure and lacking in proper sanitary and drinking water facilities.

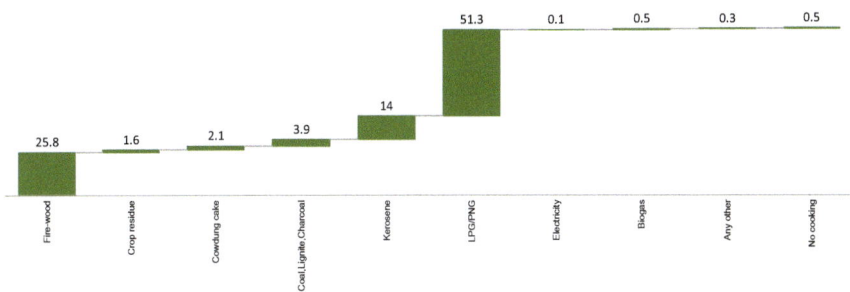

Fig. 3.17 Fuel use in slums (%) (*Source* https://censusindia.gov.in)

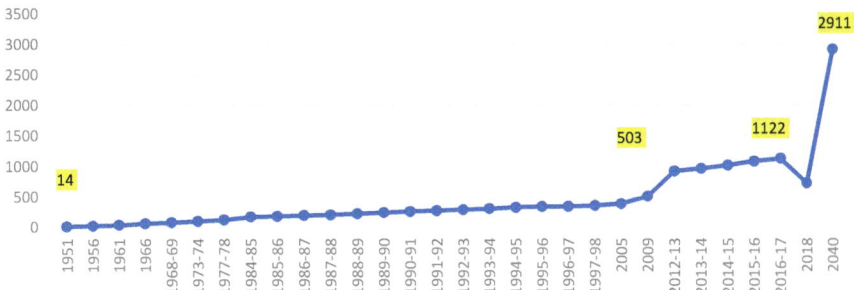

Fig. 3.18 Per capita consumption (kilowatt-hour/annum) (*Sources* Planning Commission 1953, 1956, 1961, 1970, 2001; World Bank 2013, 2017; Lok Sabha 2019b; Aayog 2017)

in 2011.[5] The main sources of lighting in these slum households were electricity (90.5%) and kerosene (8.2%). Besides, the inhabitants use substantial quantity of fuelwood (25.8%) for cooking (Fig. 3.17).

As population grows and townships expand, millions living in slums will add a new dimension to energy access and security.

India has been moving at snail's pace in improving energy generation and supply to its citizens. For more than five decades starting 1951, our per capita consumption (Fig. 3.18) was less than 500 kilowatt-hour per annum (Planning Commission 1953, 1956, 1961, 1970, 2001; World Bank 2013, 2017; Lok Sabha 2019b; Aayog 2017). This went up gradually from 503 kilowatt-hour/per capita/per annum in 2009 to 1122 in 2017. In the same year, UK and USA were consuming electricity @5130 and 12,973 kilowatt-hour/per capita/per annum, respectively. Presuming that there will not be any impediments to electricity generation, India will still find it difficult to come closer to these countries in next 20 years and the target of 2911 kilowatt-hour/per capita/per annum seems unattainable. At best, one should aim for 1700–1800 kilowatt-hour/per capita/per annum by 2040.

---

[5]Source: https://censusindia.gov.in.

3.5 Impediments to Electricity Generation

3.5.1 Hydropower

The total installed capacity (Fig. 3.19) of hydroelectric power plants in India was 44,478 megawatts (13.6% of the total installed capacity) in March 2017 despite the fact that country has enough potential to double its hydropower to 84,044 megawatts from 845 identified hydroelectric projects and provide 600 billion units per year. The state of Arunachal Pradesh alone has a potential of 50,000 megawatts of hydropower but only 98 megawatts was developed till March 2016. This sector has been slow in delivering power since construction of dams is a time-consuming process right from conception to power generation. And there are reasons why hydropower sector has lagged. Issues such as technical feasibility, finances, public opposition and demonstrations, water and land rights, environmental clearances, and legal cases are required to be resolved to exploit full potential. There are 37 new hydropower projects (above 25 megawatts capacity and with a cumulative installed capacity of 12,178 megawatts) that are under currently construction in different states, but the progress is slow due to financial and other constraints (Lok Sabha 2019c).

At present, 21 states have the capacity to produce electricity through hydropower stations with an average of 1.56 gigawatt per state (Fig. 3.20). Only two states (Karnataka and Maharashtra) have capacity above 3 gigawatts. Orissa, Tamil Nadu, Telangana, Punjab, and Himachal Pradesh have capacity between 2 and 3 gigawatt and rest are below 2 gigawatts.

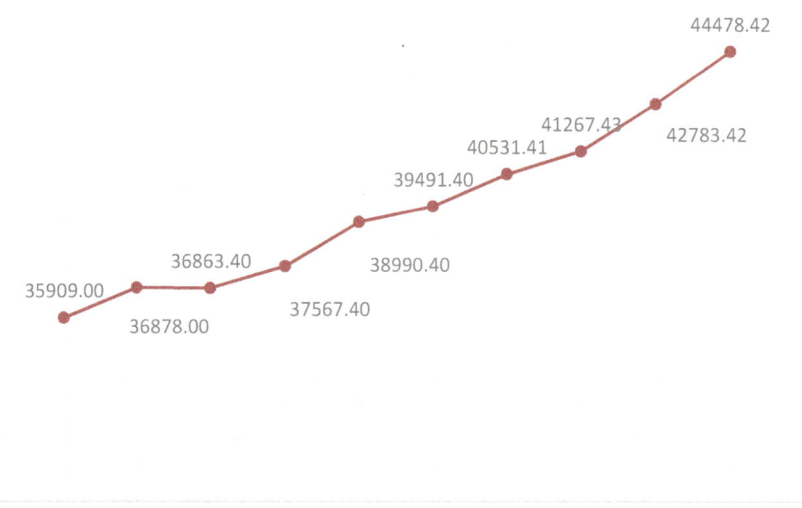

Fig. 3.19 Growth in electricity generation capacity (MW) hydro (*Source* Central Electricity Authority https://www.cea.nic.in)

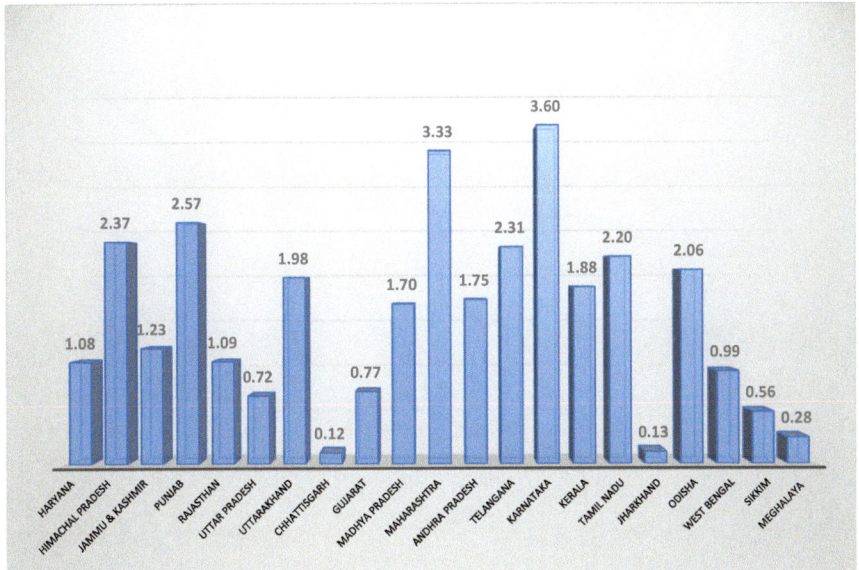

Fig. 3.20 States producing hydropower (GW) (2017) (*Source* Central Electricity Authority https://www.cea.nic.in)

Being a source of clean energy, hydroelectricity, if developed prudently and in an environmentally friendly manner, can provide multiple benefits such as water management and flood control, irrigation, and drinking water. In fact, hydropower generation was given priority over thermal and nuclear during the early period of planning after independence because flowing water is an inexhaustible resource unlike coal, oil, natural gas, and nuclear fuels which are environmentally unacceptable. Unfortunately, high cost, long-term debts, and consequent difficulties in financing medium and large hydro projects has impeded the progress of hydropower generation in India.

3.5.2 Nuclear

There are 21 nuclear reactors in the country with an installed capacity of 6780 megawatts (2.1% of the total installed capacity) (Fig. 3.21). On average, each reactor has a capacity of 323 megawatts which cannot be considered satisfactory as compared to developed nations. Moreover, of the 21 reactors, one is non-functional since 2004 (one unit of Rajasthan Atomic Power Plant, Kota) due to repairs, 13 reactors (installed capacity 4280 megawatts) use imported fuel and are under IAEA safeguard, and the rest (installed capacity 2400 megawatts) use domestic fuel. In order to ramp up power generation, construction of nine more reactors is in progress, and two more reactors

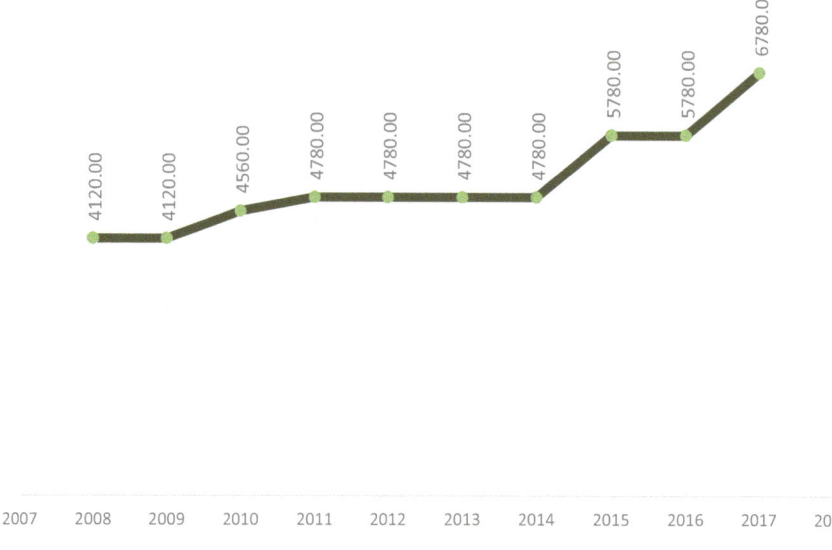

Fig. 3.21 Installed electricity generation capacity (MW) nuclear (*Source* Central Electricity Authority https://www.cea.nic.in)

(1000 megawatts each) have been approved which will enhance the nuclear capacity to 22,480 megawatts by 2030.

Considering the fact that nuclear energy has several issues and concerns including safety, waste disposal, assured fuel supply, public antipathy, and cost, it may not be a viable option in the long run. Globally, the quantity of nuclear spent fuel (Box 2) has, over the years, increased enormously and much of that has been stored temporarily at the site of plant itself. The Fukushima experience is a significant one that calls for safe disposal of nuclear waste at the earliest possible. To sum up, use of nuclear energy poses serious risks including the following:

1. Safety issues such as meltdown, explosions, malfunction, electrical error, corrosion of reactor and overheating in power stations—there have been large number of cases world-over despite all precautions.
2. Risk of radiations affecting plants, animals, and human beings.
3. Vulnerability of nuclear installations to:

   a. Enemy or terrorist attack.
   b. Theft of nuclear fuel and waste.
   c. Earthquake, tsunami, cyclone, and other natural disasters.

4. Decontamination, restoration, and reconstruction of impacted area require massive financial, technical, and human effort in the disaster-affected sites.
5. Safety of decommissioning process and the risk thereafter.

The factors listed above have motivated and/or coerced many developed countries to phase out nuclear power plants in favor of renewable energy. The current status is

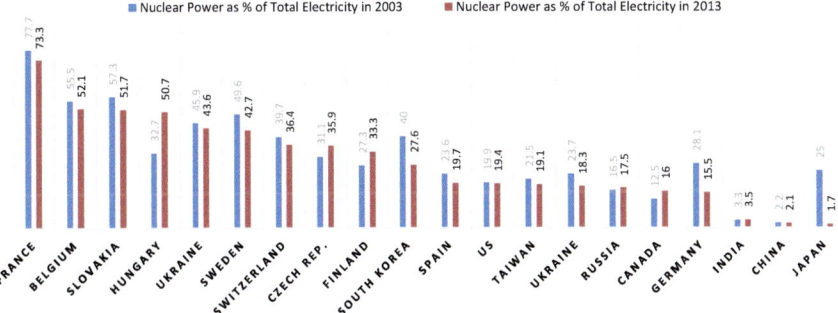

Fig. 3.22 Nuclear energy component as % of total electricity (*Source* World Nuclear Association 2014)

that Germany, that once produced 20% of its power through nuclear plants, has shut down 8 of its 17 nuclear plants and has promised to close all by 2022. France has decided to cut its share of nuclear energy generation from 75 to 50% in next 10 years; Sweden has decided to shift from nuclear energy to wind farms; Since Fukushima, all but two of Japan's 43 reactors have been suspended. Currently, only 394 nuclear plants are in operation, down 37 from 431 in the pre-Fukushima phase in 2010. A comparative analysis of variation in nuclear energy production by several countries between 2003 and 2013 (Fig. 3.22) reveals that most of the western European nations as well as USA and Canada have reduced their share of nuclear energy in total energy basket while others including India, Hungary, and Russia have increased their share (World Nuclear Association 2014).

> **Box - 2**
>
> 436 power generating nuclear reactors in 30 countries have accumulated hundreds of thousands of spent fuels and continue to increase by 10000 tons a year. Much of this is held in tanks, ponds, and other storage units. These wastes are highly radioactive. For example, plutonium 239 with a half-life of 24000 years can result in major catastrophe in the event of an accident involving these plants.
>
> [Mason, 2013]

Looking at the overall trend and public perception about nuclear energy worldwide, there are greater chances that most of the countries will curb and eventually phase out nuclear energy in favor of renewables. In any case India will not lose much as compared to many other nations (Fig. 3.23) and it will be a wise decision to shift to other alternatives at the earliest possible.

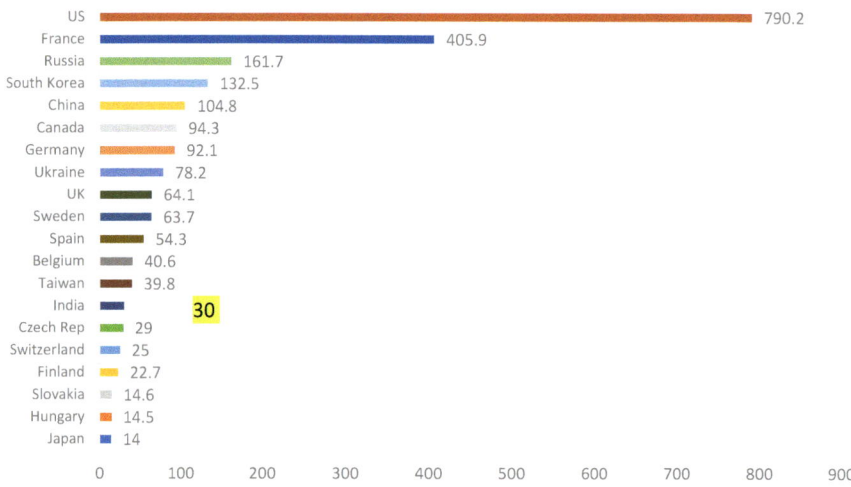

Fig. 3.23   Nuclear electricity production in 2013 (in TWh) (World Nuclear Association 2014)

3.5.3   Thermal

There are two main sources of thermal power generation in India—coal and natural gas. Figure 3.24 provides a performance overview of all states generating thermal energy.

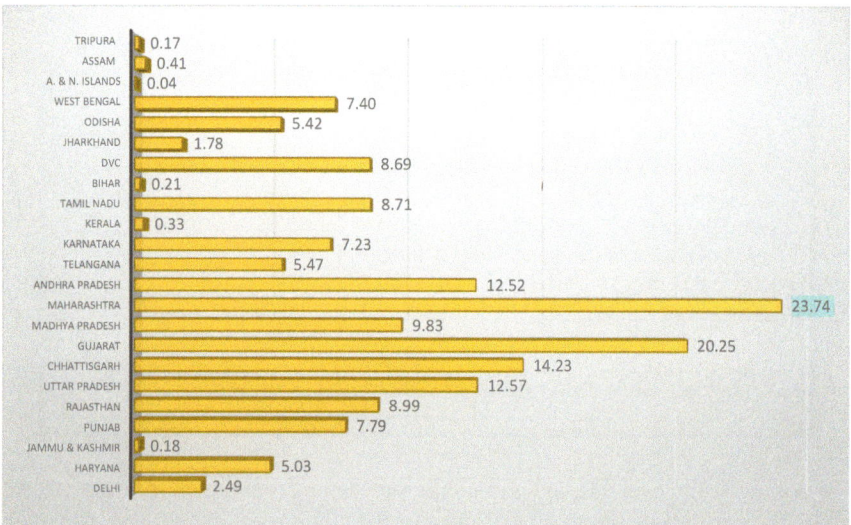

Fig. 3.24   States producing thermal power (GW) (2017) (*Source* Central Electricity Authority https://www.cea.nic.in)

3.5.4 Gas

Energy experts term the current period as 'golden age of gas' for the specific reason that fast-rising demand particularly from emerging economies for power generation and transport is the best approach to displace coal by 2040 if not earlier. Millions of tons of new capacity are coming on stream. Natural gas is undoubtedly one of the best alternatives to coal-based thermal power plants (in terms of carbon dioxide reduction). It is the second most used fuel in power generation having 8% of India's installed capacity. Unfortunately, the known natural gas reserves are only about 0.7% and the production and supply has not kept pace with the growing demand (Fig. 3.25) of natural gas in the country, including power sector. The installed capacity of gas-based power in India was 26,167 megawatts (8% of the total installed capacity) in March 2017.

Globally, India is way behind in using natural gas for electricity production with only 109 TWh in 2013 (Fig. 3.26) as compared to USA which was then producing almost ten times more electricity from natural gas.

India's annual gas production is expected to reach 90 billion cubic meters in 2040, most of which will be contributed by offshore basins, followed by onshore coalbed methane and possibly some shale gas as well. The balance requirement of 80 billion cubic meters will be met from imports. There are uncertainties in availability and indigenous production as most discoveries have been made in deep waters (depth varying between 700 and 1700 meters which is technically challenging). As a result, India will continue to meet its shortfall through imports of natural gas primarily in the form of liquified natural gas (LNG) or via pipeline from countries like Turkmenistan and Iran. In terms of cost, natural gas is too expensive vis-à-vis imported coal and therefore gas will have a limited role in India's power generation (Fig. 3.27) till such time coal is available and environmental concerns are overlooked.

Fig. 3.25 Trends in consumption of natural gas in India (billion cubic meters) (*Source* Energy Statistics 2018)

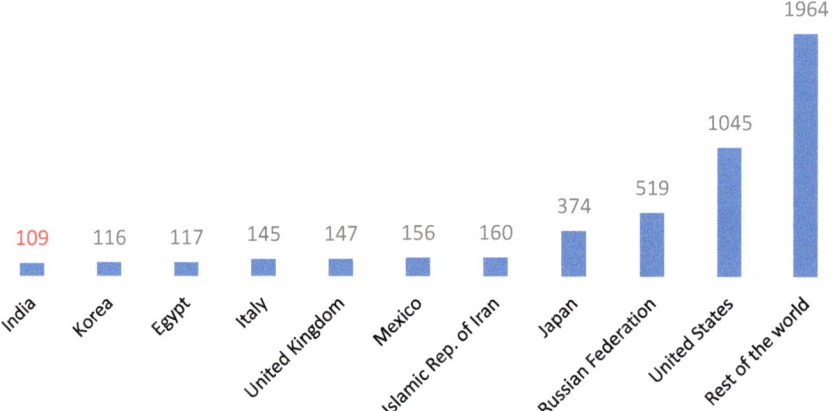

Fig. 3.26 Electricity production from natural gas—2013 (in TWh) (IEA 2013)

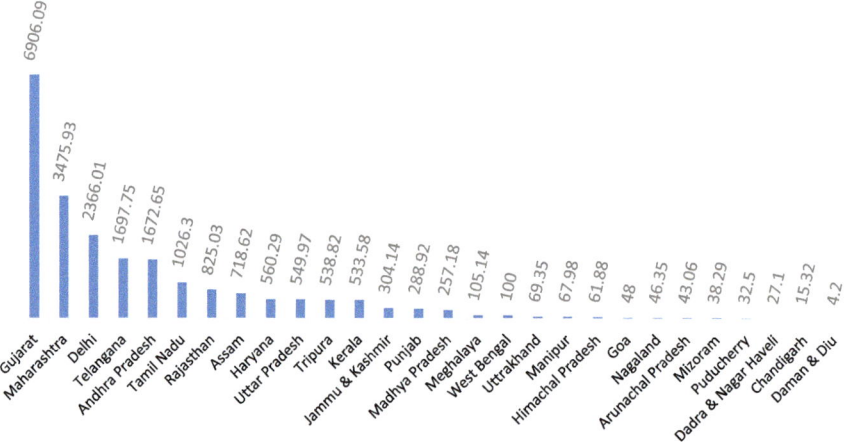

Fig. 3.27 Installed capacity of gas-based power plants (MW) as on 31-03-2015 (*Source* 'Ministry of New and Renewable Sources, Government of India'. See: https://mnre.gov.in)

Given the uncertainty of domestic gas production and availability, the share of gas-based generation is unlikely to increase significantly in future though it could have a critical role to play in meeting peak power demand and/or variability due to increased penetration of renewables.

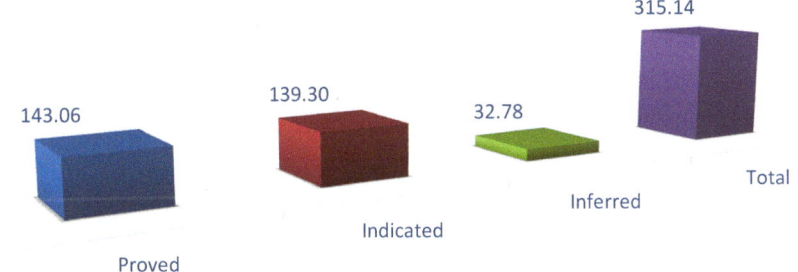

Fig. 3.28 Estimated reserves of coal in India (Billion tons) as on 31.03.2017 (*Source* Energy Statistics Yearbook 2013)

3.5.5 *Coal*

India has around 12% of the hard coal reserves of the world including large amount of lignite deposits. There are different versions of the estimate on coal reserves in India. According to the [6]World Energy Outlook Report of 2015, the total proven coal reserve in India is 87 billion tons which is almost equivalent to 140 years of output at current level (India Energy Outlook 2015). 95% of this reserve is in the form of hard coal (steam and coking coal) mostly bituminous with relatively low moisture but high-ash content and the rest 5% is lignite. However, the Government of India (NITI Aayog and Ministry of Coal) has estimated that the known levels of proven coal reserves in India range between 138 and 143 billion tons (Fig. 3.28) and this may only be able to support an annual peak production of 1.2 and 1.3 billion tons till 2037, with a gradual decline thereafter.

Coal is and will remain the backbone of Indian power system at least till the middle of this century. India is among the world's five countries that produces electricity from coal (Fig. 3.29). The total installed capacity of coal-based power plants was 192 gigawatts (192,162.88 megawatts) in March 2017 which was almost 58.8% of the total installed capacity of the country. From 192 gigawatts in 2017 the coal-based generation capacity is expected to grow to 330–441 gigawatts by 2040.[7] This is likely to translate into a coal demand of 1.1–1.4 billion tons[8] annually. Unless other sources of power generation (nuclear, hydro and renewable) follow an aggressive and determined path, there is every likelihood of coal-based generation capacity to exceed 700 gigawatts. The large planned new coal based thermal capacity is likely to put pressure on coal resources and domestic coal supplies may plateau by the year 2035. Besides, there are other serious issues with coal:

---

[6]Explanation—In order to provide a consistent basis for modeling, data from the Bundesanstalt für Geowissenschaften und Rohstoffe (BGR) on reserves and resources are used for all countries in the World Energy Outlook. The data differs from the Indian coal ministry's Coal Inventory of India report (which states 307 billion tons of coal resources and 132 billion tons of coal reserves).

[7]Ref: India Energy Outlook. World Energy Outlook Special Report 2015, IEA France.

[8]1.4 billion tons coal will release around 4 billion tons of carbon dioxide.

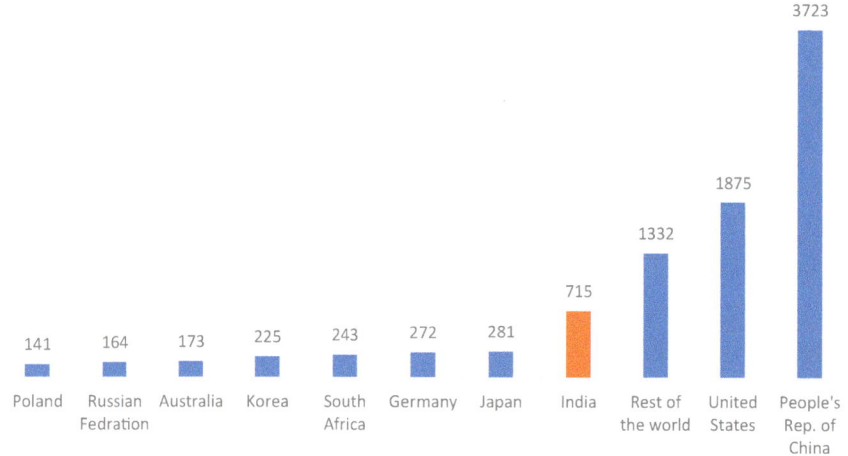

Fig. 3.29 Electricity production from coal (in TWh) (*Source* Energy Statistics Yearbook 2013)

1. A large number of existing coal-based thermal power plants are environmentally insensitive and use inefficient and low-cost technology. Most of the current technology is based on sub critical boilers where large areas of land and substantial water is needed for ash disposal which also contains significantly large quantum of mercury. In order to reduce the environmental impacts, two major technological interventions have to be introduced at a rapid pace—washing of coal with minimum or no water and gasification of coal through clean coal technologies such as supercritical, ultra-super critical and IGCC (Integrated Gasification Combined Cycle) technology irrespective of financial implications (Fig. 3.30) lest India is castigated for encouraging global warming.

2. Coal in India is available at a depth of up to 300 meters and can be exploited through surface mining. It has been estimated that on average coal-based power generation requires 247 hectares for each gigawatt and between 1545 and 54,100 km² area (Fig. 3.31) will be required till 2050 depending upon how aggressively

	Coal consumption (kg/ kWh)	2012		2047	
		Low estimate	High estimate	Low estimate	High estimate
Sub Critical	0.74	4750	4750	6191	6879.2
Super Critical	0.61	5250	5250	6879	7567.1
Ultra Super Critical	0.53	6063	6063	7739	8942.9
IGCC	0.50	9121	9121	9983	15113.6

Fig. 3.30 Current and projected capital costs of thermal generation technologies (Rs crores/GW) (*Source* NITI Aayog 2014)

	Land requirement (sq km)	
	Minimum	Maximum
Coal based Thermal Power Plant	309	1461
Coal mining	1545	54,100

Fig. 3.31 Land requirement for coal power plants and coal mining

we rely on coal-based energy. Large reserves of coal are available in Chhattisgarh, Madhya Pradesh, Jharkhand, Orissa, West Bengal, and Telangana which are incidentally rich in natural forests. Future mining may not be possible in coal reserves that are in dense forest areas as each metric ton of coal requires somewhere between 6 and 122 hectares of forest land. Another serious issue is use of children in extraction of coal from small horizontal tunnels mines in the north-eastern states including Manipur where people use rat whole mining rather than open cast.

3. Coal mining is an inherently unsafe and hazardous profession and have social and environmental implications due to displacement, pollution, and associated impacts, including deforestation and land use change. The report of the working group on occupational safety and health for the 12th Five-Year Plan (Planning Commission 2015) mentioned that coal mining is 'recognized as one of the most hazardous peacetime occupation'. Moreover, most of the existing and proposed coal-based thermal power plants are concentrated close to coal-producing areas that are already critically polluted and/or water stressed. Adding more thermal power plants in such stressed ecosystems may not be desirable from social and environmental angle.

4. Indian coal is low in calorific value and high in ash content (75% of indigenous coal has more than 30% ash), meaning thereby that more coal is consumed to generate each unit of electricity. This also means releasing extra quantity of carbon dioxide into the atmosphere adding to global warming. Compared to this, the ash content of coal traded on the international market rarely exceeds 15%. Most of the ash in Indian coal is inherent ash (small particles of mineral matter that are embedded in the combustible part of the coal) that cannot be easily removed. Most of the coal currently produced in India falls in a range of 3500 kilocalories per kilogram (kcal/kilogram) to 5000 kcal/kilogram. This is markedly lower than the average heat content of coals typically found in other large producing countries, such as China, USA, or Russia (India Energy Outlook 2015).

The National Electricity Plan 2018 provides a glimmer of hope for the climate scientists (Box 3A and 3B). The plan mentions that by 2021–2022, the coal-based generation in India will be around 63% (of the total energy generated from all sources) and the same will gradually taper off to 55.7% by 2026–2027. Based on the assessment of different sources of energy, it appears that India has the following options:

i.   A pathway for all energy supply sectors in which case the total energy-related GHG emissions in 2047 are likely to be around 10,848 MT of CO_2 equivalent, of which thermal power generation would contribute to about 3002 MT CO_2 equivalent or 27.6%; or

ii. A pathway with preference for thermal generation over renewables in which case the total energy-related GHG emissions would be around 11,342 MT CO_2 equivalent of which thermal power generation would contribute to about 3495 MT CO_2 equivalent or 30.8%; or

iii. A pathway with preference for renewable energy generation over thermal in which case the total energy-related GHG emissions would be around 9882 MT CO_2 equivalent of which thermal power generation would contribute to about 20.6%.

Box 3A - Carbon di Oxide Emission from Coal

Carbon in coal is the principal source of heat (generates 14500 BTU energy per pound) along with hydrogen, oxygen and Sulphur. Hydrogen generates heat (@ 62000 BTU per pound) but since most of the hydrogen combines with oxygen to form water, only a tiny fraction is available for heat generation. Oxygen in coal oxidizes carbon and therefore high proportion of oxygen in coal lowers its heating value. As far as sulfur is concerned, its heating value is insignificant (@ 4000 BTU per pound) and the sulfur content of coal in general averages 1-2 percent.

The proportion of carbon in coal varies from around 60% in lignite to 80% in anthracite. Carbon dioxide from coal is produced when atom of carbon (atomic weight -12) combines with two atoms of oxygen (atomic weight — 16). Complete combustion of 2000 pounds (= 907 Kgs) of coal will generate 5720 pounds (= 2595 Kgs) of carbon di oxide. In other words, one kg of coal will produce 2.86 kgs of carbon di oxide.

[Source- Hong, 1994]

Box 3B - Coal Drudgery

1. An average Indian coal miner produces less than 2500 tons of coal per year, whereas an Indonesian miner is at least 50% more productive. A coal miner in China produces more than 5000 tons per year and an Australian worker up to 13000 tons per year on average.

2. In India, poor wages of the mines lead to a higher labor-intensity than elsewhere in the world.

3. Since the early 2000, production of high (**more** than 4200 kcal/kg) energy coal has remained stagnant while medium (**less** than 4200 kcal/kg) energy coal production has more than doubled meaning that miners in India have to extract around 1.5 tons of coal to get the same amount of energy as that contained in one ton of Australian coal;

4. Production costs for coal in India fall in a wide range, with some large open-cast mines producing coal for less than $15 per ton, while other small high-cost underground mines have costs in excess of $150 per ton.

5. Ash disposal is also a problem as fly ash utilization (e.g. in the cement or brick industries) absorbs only around 60% of the total yield. The balance adds to pollution.

[Source: Outlook, India Energy, 2015]

3.5.6 Oil

India, being poor in oil resources as compared to many countries (Fig. 3.32), meets its indigenous crude oil primarily from three onshore states (Gujarat, Assam, and Rajasthan) and one offshore (Mumbai High). Less indigenous production and high demand (Fig. 3.33) forces large-scale imports which is a huge drain on India's foreign exchange reserves. In 2014, India imported oil and gas worth USD 110 billion (5.3% of GDP) which is expected to balloon to more than USD 300 billion in 2030 and USD 480 billion in 2040. India's oil production is expected to decline overtime to around 700 kb/d with a consequent rise in net oil imports to 9.3 mb/d by 2040.[9] Any unforeseen global situation like war or disease outbreak may lead to perilous economy.

---

[9]Source—India Energy Outlook (2015).

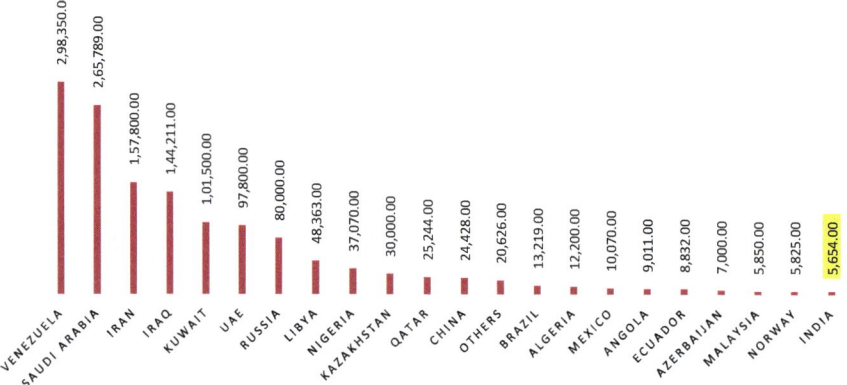

Fig. 3.32   Oil reserves (million barrels) (*Source* Energy Statistics Yearbook 2013)

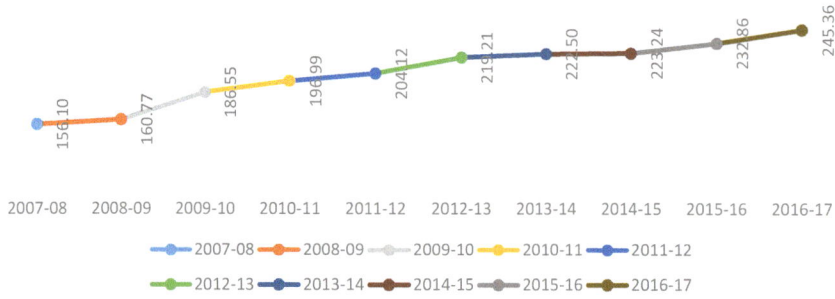

Fig. 3.33   Trends in consumption of crude oil in India (Million Metric Tons) (*Source* Energy Statistics Yearbook 2013)

Notwithstanding its modest oil reserves, India has over the years improved its refining capacity and is the fourth largest refining country in the world behind USA, China, and Russia.

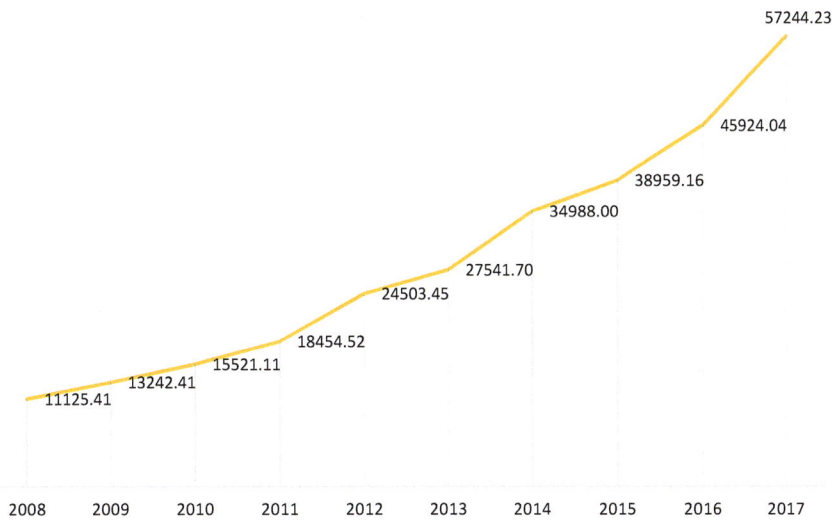

Fig. 3.34 Growth in renewable electricity generation capacity (MW). (*Source* Central Electricity Authority https://www.cea.nic.in)

3.5.7 Renewables

India has a very ambitious plan for switching over to renewable sources of energy. From 57 gigawatts (57,244 megawatts) in 2017 (17.5% of total installed capacity), the target for the year 2022 is 175 gigawatts and thereafter 597–710 gigawatts by 2040. Government is promoting renewable energy, specifically solar, as India is one of the best recipients of solar energy due to its location in the solar belt and has vast solar potential of 749 gigawatts for power generation. Also, India has substantial wind potential of 103 gigawatts due to its long coastline. Decadal growth in renewable energy sector has been phenomenal from almost 11,000 megawatts in 2008 to 57,000 megawatts in 2017. The details are provided in Fig. 3.34.

Even though there is tremendous scope for expansion in renewable sector, many states have pulled the growth downwards by performing much below the expectations. Sixteen states/UTs have installed capacity below 100 megawatts and most of these are in the north-eastern part of the country having tremendous potential for renewable energy. Eleven states/UTs have capacity between 101 and 1000 megawatts and only ten states have shown good progress over the years with installed capacity of more than 1000 megawatts. The details are provided in Figs. 3.35, 3.36 and 3.37.

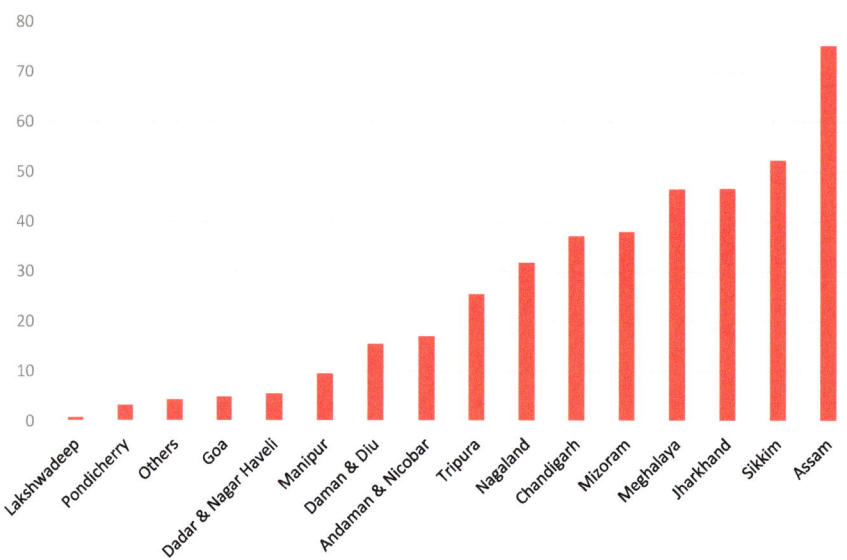

Fig. 3.35 State-wise installed capacity (less than 100 megawatt) of grid interactive renewable power (small hydro, wind, bio, and solar) as on 31.08.2019 (Ministry of New and Renewable Sources, Government of India'. See: https://mnre.gov.in)

Unfortunately, only seven states make good use of wind energy and five states generate appreciable quantity of biopower, whereas most have developed solar power. A comparison (Fig. 3.38) of installed capacity of renewable energy and its potential reveals that almost all states lag in their commitment toward less carbon energy systems. The performance of each state must be closely monitored to ascertain the reasons for appalling execution of renewable energy. At this pace India will find it difficult to fulfil its commitment to UNFCCC.

The target for 2022 is to have 175,000 megawatts (renewable energy capacity (comprising 100,000 megawatts solar, 60,000 megawatts wind, 10,000 megawatts biomass, and 5,000 megawatts small hydro). In 2015, the installed capacity of renewable was mere 37,400 megawatts (24,400 megawatts of installed wind energy and 4300 megawatts of solar energy).

Renewable is the future of energy and it will be inexpensive (and affordable by all), efficient, easy to maintain, easy to generate, and zero carbon emission.[10] The most

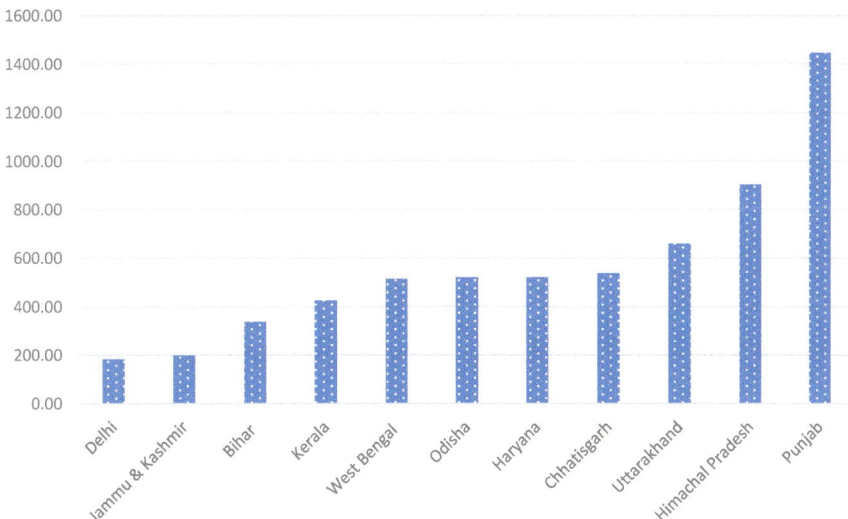

Fig. 3.36 State-wise installed capacity (between 100 and 1000 megawatt) of grid interactive renewable power (small hydro, wind, bio, and solar) as on 31.08.2019 (*Source* 'Ministry of New and Renewable Sources, Government of India'. See: https://mnre.gov.in)

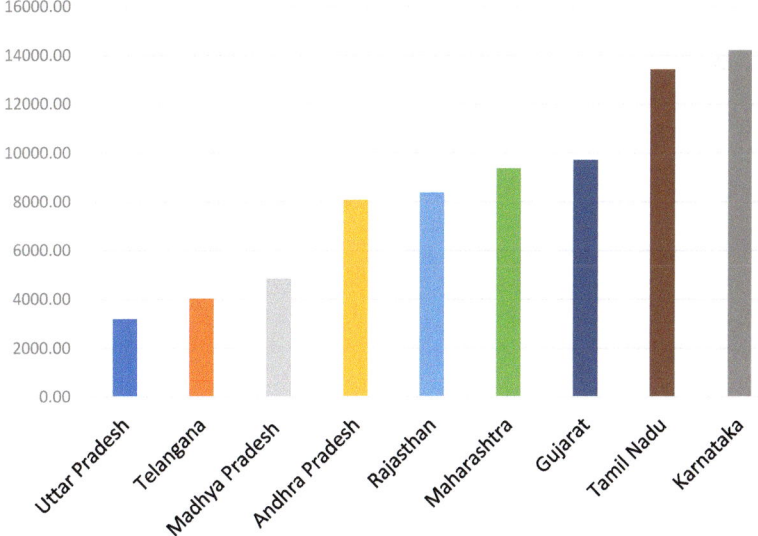

Fig. 3.37 State-wise installed capacity (more than 1000 megawatt) of grid interactive renewable power (small hydro, wind, bio, and solar) as on 31.08.2019 (*Source* 'Ministry of New and Renewable Sources, Government of India'. See: https://mnre.gov.in)

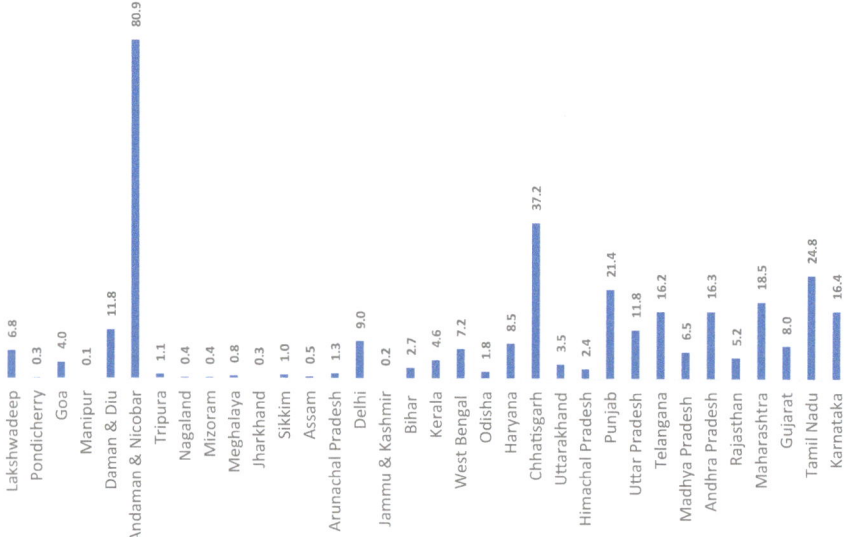

Fig. 3.38 Deficit (installed capacity of renewables versus potential of renewables (in %)) (*Source* 'Ministry of New and Renewable Sources, Government of India'. See: https://mnre.gov.in)

attractive part of renewable is that it can be used in a decentralized manner—from individual building to large complexes and in rural as well as urban areas. In fact, a household can generate its own need through solar rooftop system using photovoltaic panels. Solar lights with sensors can be used for longer duration for lighting homes. Solar collectors and heat pumps can provide low carbon heating. Solar cookers are the best substitute for replacing firewood and other organic waste. Windmills can help villagers draw water from wells. In a nutshell, next 10–20 years will be golden period for Indians to switch over from coal to green electricity. Renewable energy is the best bet for 'Net Zero Energy – Zero Carbon' housing that currently consumes nearly 40% of energy for lighting, heating, cooling, and ventilation. Simple changes such as judicious placement of windows and roof shadings, natural green surroundings, high ceilings and vents, efficient house lighting, placing solar panels on roof tops will revolutionize energy woes (Box 4) in India. Energy saved in urban areas can be distributed to villages to help them set up small and medium industries.

---

[10]Each kilowatt hour of electricity (about as much as it takes to run a dishwasher) generates over 0.5 kilogram of carbon dioxide.

Box 4 - ELECTRICITY – SHOCK AND AWE

This story was published in The Hindu newspaper on August 27, 2016 under the title ēThe race to light up the last village. The report highlights massive efforts by the state to ensure that villagers get the benefits of electricity under the flagship program of rural electrification launched by the government in 2015.

Every village that has remained deprived of electricity has a painful narration wrapped under the soul and mind of its inhabitants. This pain gets aggravated when they travel/migrate to cities to get dazed by the electronic goods, sound systems, mobiles phones and similar devices. And they are overwhelmed when their homes are lit by LED bulb, fans, cooler and TV. Tanwarpura and Salam Singh Ki Basti are two villages in Barmer, Rajasthan whose residents had remained devoid of electricity connection for almost seven decades post-independence. While most of the cities enjoyed the pleasures of electrical appliances, inhabitants of these two villages were using kerosene lamps, manually operated flour mill and wood fuel. They had to spend money for travelling 25 kms to the neighboring village to get their mobile phones recharged. Electricity connectivity gave them opportunity to buy refrigerator, electric flour mill and saved them from drudgery of mobile recharge. The realization that their children will be saved from the obnoxious fumes of kerosene lamps gave them immense pleasure.

Anandpur village in Hathras district of UP has a different story. women felt happy that they were freed from swatting insects that would hover around kerosene lamps while cooking food. Resident were earlier scared of venturing out after dusk due to large number of snakes and presence of LED bulb boosted their confidence. This story is an indicator of shock and awe that we as humans experience with improvement in living condition.

An attempt has been made below to explain different stages of device acquisition by urban India. Rural India is way behind and may take several decades to reach the current level of urban acquisition.

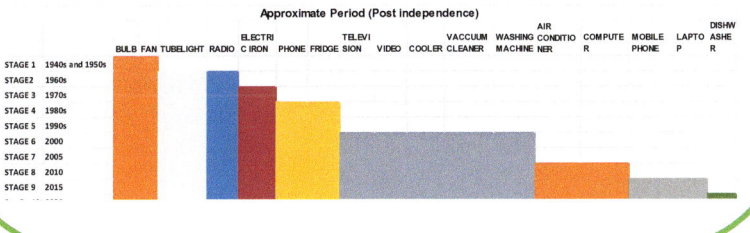

3.6 Biomass Fuels: A Threat to India's Forests, Soil, and Human Health

The working group on energy policy set up by the Planning Commission in 1979 (Planning Commission 1979) reported that India consumed 86.3 million tons of firewood, 26.4 million tons of agricultural waste, and 46.4 million tons animal dung in

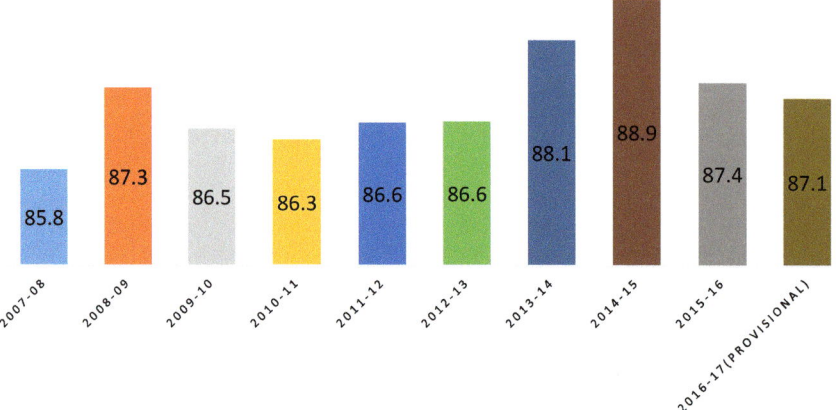

Fig. 3.39 Domestic distribution of LPG (percent of total LPG use) (*Source* Energy Statistics Yearbook 2013)

1953–1954 (human population 379 million). Rather than decline, the consumption[11] showed an upward trend in 1999–2000 with 191.6 million tons of firewood, 59 million tons of agricultural waste, and 105 million tons of animal dung (human population 1028 million). The NSS report 567 (68th round, 2011–2012)[12] showed that there was a decline in rural solid biomass-based cooking between 1999–2000 and 2011–2012 by 8.2% while LPG uptake increased from 5.4 to 15%. The report also mentions that only 1% rural and 6% urban population were using kerosene for cooking.

Ever since independence, the great rural–urban divide in use of cooking fuels has widened. Unlike urban household, the rural ones do not have multiple choice of cooking fuel compelling them to stockpile fuelwood, agriculture waste, and dung cake as an insurance in case of volatility in LPG supply.[13] LPG in India is used primarily in domestic supply and transport sector. The total consumption of LPG has increased from about 12 million tons in 2007–2008 to about 21 million tons in 2016–2017 but the proportion of domestic use (Fig. 3.39) has remained almost constant.

Since LPG supply is largely dependent on imports, absence of assured market and distorted supply chain will keep the consumer especially in rural areas depend on biomass fuel for several years or till such time people have assured and affordable substitute/s. The other non-biological source of fuel used by poor families both in rural and urban India is kerosene. More than 90% of available kerosene is used for domestic purpose and the total consumption of kerosene has declined from about 9

---

[11] As reported in seventh plan document.

[12] Ministry of Statistics and Programme Implementation, Government of India. "Energy sources of Indian households for cooking and lighting, NSS 68th round, July 2011–June 2012." (2015).

[13] As per India Statistics 2018 published by Central Statistical Office, Ministry of Statistics and Program Implementation, Government of India, the domestic production of LPG and kerosene in 2015–2016 was 11.33 and 7.50 million tons, respectively.

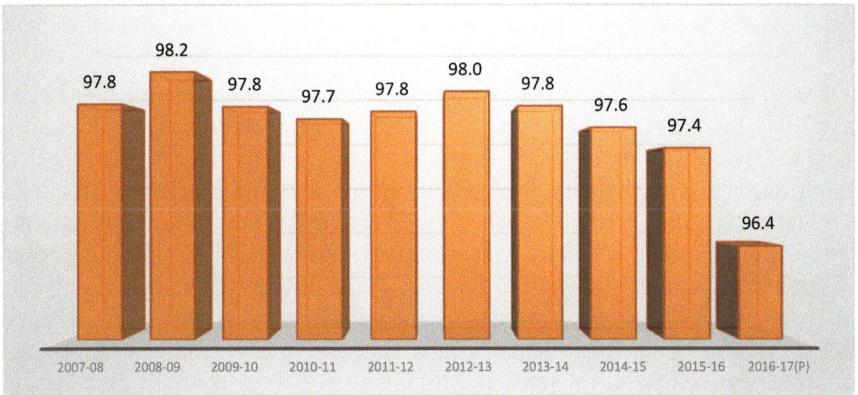

Fig. 3.40 Domestic consumption of kerosene as percent of total source (Energy Statistics 2018)

million tons in 2007–2008 to about 5.5 million tons in 2016–2017. But as in the case of LPG, the proportion of kerosene (Fig. 3.40) in domestic use has remained almost constant.

Unfortunately, there is no comprehensive study on the consumption of biomass fuel in different states of the country nor on the substitution rate of biomass fuel to non-biomass fuel. There were 840 million Indians in 2015 whose major source of cooking fuel was fuelwood[14] and this number may see a marginal decline in 2030, when out of 2.7 billion[15] people relying on plant-based energy, 782 million will be from India.

A special report prepared by International Energy Agency in 2015 and titled India Energy Outlook mentions the following:

Today more than 70% of energy used in households in India is for cooking (whereas cooking constitutes less than 5% of residential energy demand in OECD countries). Two-thirds of the Indian population rely on solid biomass as their cooking fuel (Government of India 2012), due to the lack of options that are similarly available and affordable; the low efficiency of this cooking method, compared with LPG or electric stoves, pushes up the share of solid biomass in cooking energy demand to more than 85%. Changes in the fuels used for cooking account for some of the main changes in residential energy demand over the period to 2040, alongside fuel switching for lighting purposes from kerosene (mostly in rural areas) to electricity, and rising electricity consumption to meet large increases in demand for cooling equipment and appliances. There is not much call for space heating in India, as daytime temperatures in its most populated areas are on average higher than 20 °C.

In most rural areas in India, it is a challenge to displace solid biomass as the dominant fuel for cooking. Biomass scarcity is not yet at the level at which it forces a transition to other fuels and although LPG is promoted as an alternative and each household is entitled to buy 12 LPG cylinders per year and to receive the related subsidies as direct payments to their bank account, distribution networks for LPG are limited in rural areas and, even with the subsidy, the cost can deter the poorest households. Biogas seems a promising avenue for

---

[14]India Energy Outlook (2015).

[15]Ref WEO (2006).

India (based on ample agricultural residues) and there have been long-standing efforts to promote it, but less than 1% of households use biogas as their primary cooking fuel.

Use of biomass energy defeats the very purpose of human development. Since most of the biomass using households are poor, they rely on low-quality cookstoves, cooking utensils, and live in poorly ventilated housing which exacerbates the negative health impact, as there is incomplete combustion and non-dissipation of smoke. Burning of biomass fuels produces high levels of carbon monoxide, hydrocarbons, and particulate matter, which in turn have disastrous impact especially on the health of women (excessive contact) and children (susceptibility). Children exposed to indoor air pollution are two to three times more likely to catch pneumonia, which is one of the world's leading killers of young children. In addition, there is evidence to link indoor smoke to low birth weight, infant mortality, tuberculosis, cataracts, and asthma. The effects of exposure to indoor air pollution depend on the source of pollution (fuel and stove type), pollutant dispersal (ventilation), and exposure time (how much of time household members spend indoors).

The burden of collection of fuelwood lies with women and children with significant opportunity cost in terms of loss of education and income generating activities. Many children, especially girls, are withdrawn from school to attend to domestic chores related to biomass use, reducing their literacy and restricting their economic opportunities.

It is almost impractical to ascertain the exact nature of biomass consumption in household sector in a huge country like India. There are enormous variations in the level of consumption and type of fuel used for cooking and heating. While most of the rural families are aware of the sufferings and consequences of biomass burning within their four walls, factors like poverty, willingness to pay, uncertainty in availability of alternatives compel them to resort to biomass fuels. There have been instances in big cities and towns of poor people burning tyres, polythene sheets, plastic bottles, and so on for warmth during winters. Since poverty is one of the major factors, the impact of pollutants is highest among people with low income and minimal savings.

3.7 Overcoming Impediments to Sustained Energy Supply

Energy plays a significant role in human development and countries with low HDI scores have low primary energy consumption per capita.[16] In countries where more than 75% of the population receives less than $2 per day income, energy consumption (biomass plus commercial) is around 0.4 tons of oil equivalent (toe) per capita per annum. As the percentage of poverty indicator falls to 40–75%, energy consumption rises to about 0.8 tons of oil equivalent (toe) per capita per annum, and at 5–40% fall in poverty indicators, it rises to over 1.5 tons of oil equivalent (toe) per capita per annum. If a country must achieve high HDI score, it should consume more than 2 tons of oil equivalent per capita per annum energy consumption.

---

[16]Source—Srivastav (2019).

This is a trap for many countries including India. In the race for achieving high HDI score and in the absence of economically viable alternatives, India will continue to use coal irrespective of the environmental and health consequences. The chances of reversing the trends appear bleak as it poses a formidable challenge between economic growth, energy security, and climate mitigation. Huge upfront investment is required in the coming years for:

(a) Switchover from sub-critical thermal power plants to super-critical, ultra-super-critical and IGCC.
(b) Complete small, medium, and large hydropower plants on time.
(c) Enhance the capacity of renewable energy.
(d) Complete substitution of biomass fuels in rural and forest areas.

While the current policies indicate that India is inclined to follow low carbon path, but the data indicates otherwise, that is, sluggish transition toward low carbon trajectory. India's absolute emissions are expected to grow much beyond 2030 and the impact of climate-related events will be much bigger than anticipated. Consider the following:

i. India plans to increase its coal production to one billion tons by 2019, and the absolute growth in coal-generated electricity by 2030 is larger than absolute increase in renewable energy generation capacity. By 2040, India will be the second largest coal user (after China) generating 441 gigawatts[17] under business as usual scenario.

ii. India's current climate change targets are insufficient to counter the catastrophic events, which according to IPCC will worsen in future.

iii. India's clean energy finance has been inadequate. India's investment in clean energy has averaged 9 billion USD per annum between 2010 and 2015. In 2014, India invested 7.9 billion USD as compared with China's (83 billion USD) and USA (35.8 billion USD).

iv. In 2016, there were 62 million households without access to grid electricity.[18] In addition, there were around 11 million households in rural areas which were officially connected with grid but did not get adequate power supply due to poor service and outages.

v. In 2017, there were 304 million Indians without access to electricity and 500 million depended on biomass fuel for cooking.[19] For such people double-digit growth rate has no meaning and they continue their struggle for fuel collection.

vi. There are serious challenges in sustaining rural electrification. Rural poor still believe that electricity is a social welfare activity and not a serious commercial business. Government schemes can provide household connectivity to the grid but getting the cost of electricity supply will be a big challenge. That is why not many private players will be keen to invest in rural energy sector.

---

[17]NITI Aayog (2017).
[18]Watanabe et al. (2016).
[19]NITI Aayog 2017.

vii.   With vast majority of poor people, the price of electricity is a sensitive issue in India, compelling the government to keep in check the costs of power generation, transmission, and distribution. The average cost of power generation is expected to increase from around $65 per MW-hour (MWh) to over $70/MWh in 2040 (India Energy Outlook 2015).

viii. Access to clean cooking fuels in rural India—the substitution of dirty fuels such as firewood, dung, kerosene, coke, and coal to LPG, PNG, and electricity has been a nightmare. LPG may be a preferred fuel by rural consumer, but the cost of a bottle of 14.2 kilogram is prohibitive, supply is erratic, and there is hardly any home delivery network. With around 50% of LPG being imported, there is no assured supply to rural areas. Stacking of fuel being a norm in villages especially remote ones, people will continue to use multiple fuels including biomass in the coming decades. Besides electrification, one of the most critical aspect of energy is the fuel for cooking. In India, household electrification and provision of clean cooking fuel have been twin challenges, with the former having received priority over the latter. This has led to poor redressal of this issue, resulting in near 40% of our population without access to clean cooking fuel. The situation in rural areas, with a significant section of the populace below poverty line, is grim, and is changing at snail's pace from firewood, dung cake, kerosene, and coal to LPG and PNG.

ix.   The complex web of rural electrification cannot be resolved easily. Government efforts can create electricity infrastructure and provide connectivity to each permanent household in a time-bound effort. But there will still large number of migrant families (mainly laborer) who stay in makeshift houses with their families for a substantial part in a year. Besides infrastructure and connectivity, adequate supply of desired quality of power at affordable rates, providing clean and sustainable power in an efficient manner for long term is a challenge that has no easy answers.

x.   A few issues that have kept the electricity development agenda on the backfoot include:

   a.   Savings and willingness to pay: This is a big setback as a large proportion of poor population residing in villages and hamlets either have very little extra money to afford commercial electricity or they are not willing to pay. The contrarian view, however, is that the rural consumers value electricity as much as anyone else and are willing to pay reasonable tariffs. This view may be acceptable for those villages which are close to urban areas and where income levels are high. For poverty-stricken villages in economically backward states, food, fuel, health, and savings are priority over payment at commercial rates for electricity.

   b.   Unresponsive attitude: The poor rural consumer is indifferent toward electrification. Having born and brought up in dark (devoid of electricity) household, many villagers have adapted to such life and livelihood and therefore are not enthusiastic or frontrunner for change. Such persons or families would either expect commercial electricity to be provided free of cost or heavily subsidized. Or else they may illegally tap the connection, a phenomenon that many urban cities have witnessed in the past and the same continues.

c. Perception: Like other development programs of the government including construction of roads, houses, toilets, schools, water supply, and panchayat buildings, rural electrification is reckoned as a welfare activity, and not a commercial venture. The psyche of subsidy, aid, assistance, and relief (except disaster related) and the politics of vote encourages the supplier and consumer to consider rural electricity supply as a social welfare activity that hampers the recovery of cost.

d. Cost effectiveness: Solar lighting may be an effective solution for household electrification, but not cost-effective now in many cases, especially if battery back-up is also provided.

e. Duration, quality, and reliability: Providing connection is no end. Indian village being a socially complex web, issues such as maintenance of transformers, categorization and divide between below poverty line (BPL) and above poverty line (APL) will be crucial for sustenance of village electrification program.

The ongoing circumstances give rise to many such as:

1. Can India provide enough electricity per capita to be at par with developed nations? If yes, how and by which year?
2. Can India provide universal, secure, adequate, and affordable electricity to all households on $24 \times 7 \times 365$ basis by 2040?
3. If yes, will this energy be clean energy, and will India be able to meet its climate mitigation commitments?
4. If not, what will the impact on human health, education, and economy?
5. Without a continuous focus on emissions control technologies in the power sector, industry, and transport, does India face serious risk of a deterioration in urban and sub-urban air and water quality?
6. More than 800 million Indians still rely on biomass fuel for cooking and 304 million are without access to electricity. Can these figures be brought to naught and if yes do we have a deadline?
7. Can India fulfil its global commitment of ensuring renewable energy capacity to 175 gigawatts by 2022 as well as ensuring GHG emission reduction by 33–35% by 2030; If yes, how?
8. In 2017 our per capita electricity consumption was 805 kilowatt-hour/year,[20] that is, less than 3 units of electricity per person per day way below that of USA at 12,973 kilowatt-hour/year. Can we ever aspire to reach 10,000 kilowatt-hour/year?
9. Double-digit growth rate has little meaning for a family that struggles for biomass fuel collection and spends 3–4 hours per day for this purpose only.

---

[20] Source World Bank (2017).

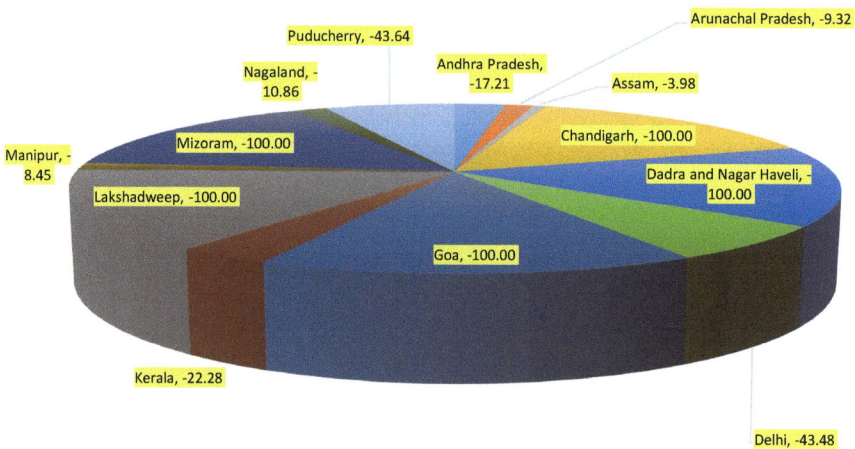

Fig. 3.41 Decline (%) in electricity generation (MWh) between 2006 and 2016 (*Source* Energy Statistics 2018) (Decline in Andhra Pradesh is due to formation of Telangana)

10. The energy demand of India is likely to go up by 2.7–3.2 times between 2012 and 2040, with the electricity component itself rising 4.5-fold. Can Indian states afford decline (Fig. 3.41) in energy production at this juncture?
11. Can sectors such as beauty and wellness, media and entertainment, leather goods and furniture continue to dominate employment sector (Fig. 3.42a, b) for next 20 years or will it necessarily change to manufacturing, software, communication, transport if India must maintain growth trajectory. If yes, do we have roadmap?

3.8   India's Domestic Energy Balance in 2040

Energy security involves ensuring uninterrupted supply of energy to support the household and commercial activities necessary for sustained economic growth. Energy supply is obviously more difficult to ensure if there is large dependence on imported energy. Overcoming the impediments in energy supply calls for action in following areas:

1. Enhancing domestic production of hydro, gas and renewable energy by resolving impediments including land availability, expediting impact assessment and requisite approvals for environment and forest and implementation of the Scheduled Tribes and Other Traditional Forest Dwellers (Recognition of Forest Rights) Act, 2006.
2. One important reason for delay in obtaining approvals is the piecemeal approach and pressure tactics used by the implementing agency. Such approaches at times prove counterproductive and lead to delays. It is, therefore, vital that a 50-year

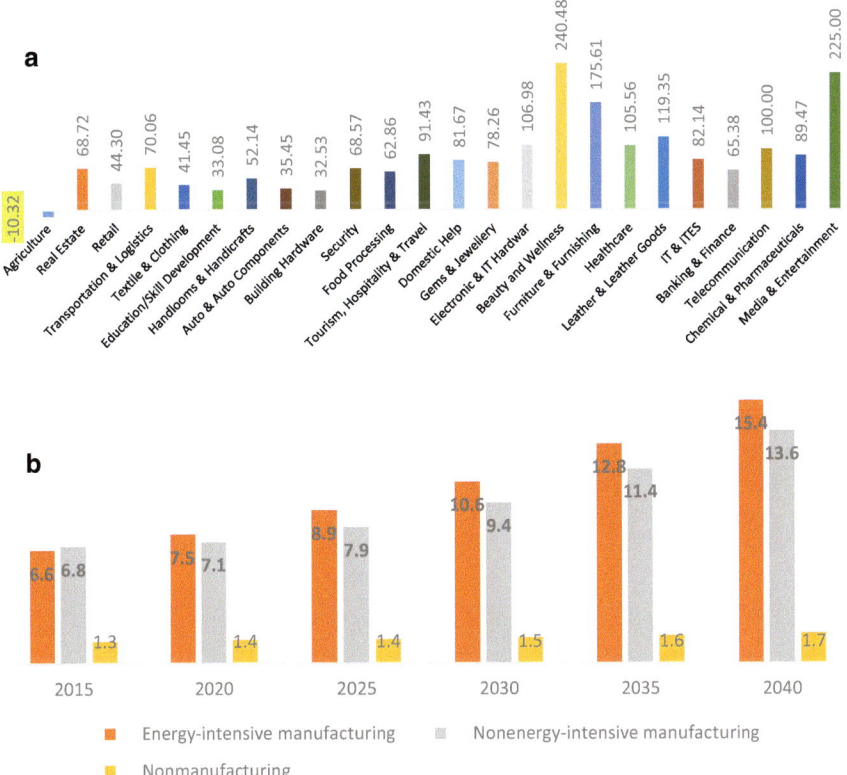

Fig. 3.42 a Percent change in employment (between 2013 and 2022) (*Source* Annual Report (2015–16) (2016). Ministry of Skill Development and Entrepreneurship. Website www.skilldeve lopment.gov.in). **b** India's industrial sector energy consumption (Quadrillion BTU), by case and sector, 2015–2040 (*Sources* International Energy Outlook 2018)

roadmap is prepared without compromising the environmental damages, use of state-of-the-art technologies and financial integrity.

3. An unambiguous legal and policy regime must be provided to attract long-term private investment including foreign investment in oil and natural gas blocks and new capacities for renewable energy. Investors should be given assurance about the profitability and stability of tax regime.

4. Public investments in renewable energy companies should be encouraged with guaranteed returns. A number of private companies that invested in renewable energy sector are currently in the red and investors have suffered huge losses. At current level of investment and foresight the share of renewable energy in total energy consumption will not be a substantive.

5. Investments in energy assets in foreign countries, especially for natural gas and petroleum should be stepped up.

6. Oil storage capacity enhancement is a prerequisite for all import-dependent nations. *The Organization for Economic Cooperation and Development (OECD)* members *have* a storage capacity of up to 90 days. But it will be useful to have more than 90 days storage to avoid balance of payment crisis at critical times.

Energy demand (Fig. 3.43) in India is projected to escalate over the next two decades with population expected to cross 1.6 billion by 2040. This will have far-reaching implications for the energy consumption as 45% population is expected to settle in urban areas largely in slums. The demand for energy use will reach 1900 million tons of oil equivalent propelled by economic consideration (expected to reach five times its current size in 2040), and to meet this demand, India needs to build more than 880 gigawatts (GW) of new power generation capacity.

All indicators point to the fact that coal will retain a central position in the energy mix, increasing its overall share in primary energy to 49% in 2040. Demand for oil will also reach 9.8 mb/d with 65% rise in transport as 260 million additional passenger cars, 185 million new two and three-wheelers and nearly 30 million new

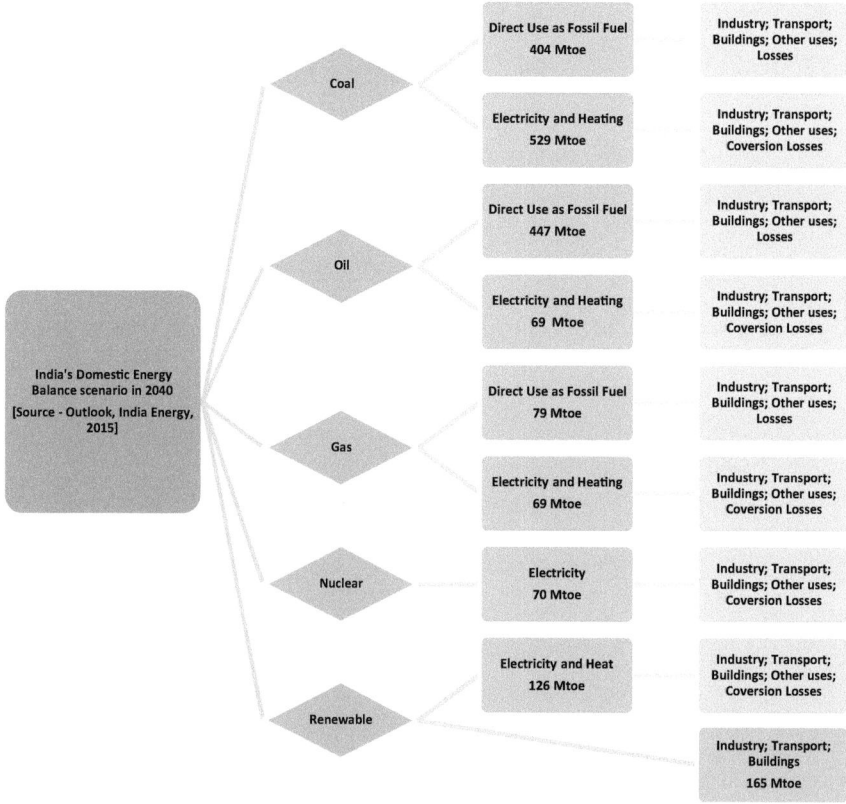

Fig. 3.43 India's Domestic Energy Balance scenario in 2040 (*Source* India Energy Outlook 2015)

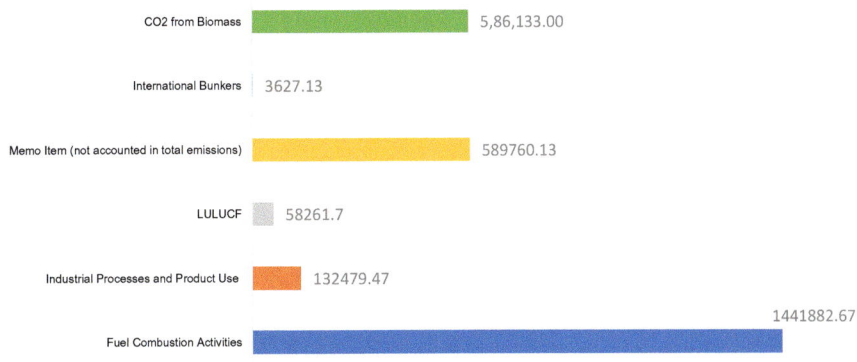

Fig. 3.44 Carbon di oxide emission (gigagram) in 2010 (*Source* India Energy Outlook 2015)

trucks and vans are expected to be added to the vehicle stock. Natural gas, though versatile and low in environmental footprint, will play a relatively minor role in the Indian energy mix due to its high price that does not allow it to replace coal or coal sooner (India Energy Outlook 2015).

The supply of adequate and affordable power has to overcome numerous internal as well as external dynamics to improve the reliability, quality, and integration with renewable energy technologies. India's high dependency on various countries for fuel procurement, technology cooperation, and import and flow of investment will be crucial. By 2020, India will be the largest importer of coal in the world and will have one-sixth of global solar photovoltaic capacity by 2040. One may not appreciate at this moment, but in all likelihood, India will have to face international criticism for spewing venom through carbon dioxide (Fig. 3.44) emission.

References

Annual Report (2015–16) Ministry of Skill Development and Entrepreneurship. www.skilldevelop ment.gov.in

Central Electricity Authority. https://www.cea.nic.in/

Chandramouli C (2011) Census of India 2011. Provisional population totals. Government of India, New Delhi, pp 409–413. https://censusindia.gov.in

District Census Handbook (DCHB) https://censusindia.gov.in/2011census/dchb/DCHB%202011-Concepts%20&%20Definitions%20Village%20and%20Town%20Directory.pdf

Energy Statistics (2018) Central Statistics Office, National Statistical Organization. Ministry of Statistics and Programme Implementation, Government of India

Energy Statistics Yearbook (2013) Statistics Division, Department of Economic and Social Affairs, United Nations, New York

Hong BD, Slatick ER (1994) Carbon dioxide emission factors for coal. Quarterly coal report 7

https://www.ddugjy.gov.in/page/definition_electrified_village

IEA (International Energy Agency) (2013) Key world energy statistics 2013

India Energy Outlook (2015) World energy outlook special report 2015. International Energy Agency. https://www.iea.org/publications/freepublications/publication/IndiaEnergyOutlook_WEO2015.pdf

Lok Sabha (2018) Question no. 1813. Govt. of India

Lok Sabha (2019a) Question no. 770. Govt. of India

Lok Sabha (2019b) Lok Sabha question 337 (Archive)

Lok Sabha (2019c) Lok Sabha question 798 (Archive)

Mason C (2013) The 2030 Spike: countdown to global catastrophe. Routledge

Ministry of New and Renewable Sources, Government of India. https://mnre.gov.in

Ministry of Power (2018) Annual report 2017–2018

Ministry of Statistics and Programme Implementation, Government of India (2012) Energy sources of Indian households for cooking and lighting, NSS 66th round, July 2009–June 2010

Ministry of Statistics and Programme Implementation, Government of India (2015) Energy sources of Indian households for cooking and lighting, NSS 68th round, July 2011–June 2012

Muralidharan K (2013) Priorities for primary education policy in India's 12th five-year plan. In: India policy forum, vol 9, no 1. National Council of Applied Economic Research, pp 1–61

NITI Aayog (2014) User guide for India's 2047 energy calculator cooking sector

NITI Aayog (2015) Government of India. India energy security scenarios 2047, version 2.0. India Energy Portal

NITI Aayog (2017) Draft national energy policy. National Institution for Transforming India, Government of India, New Delhi. https://niti.gov.in/writereaddata/files/new_initiatives/NEP-ID_276

NITI Aayog (2017) Draft national energy policy-India. NITI Aayog, Government of India, pp 1–106

Planning Commission (1953) First five year plan

Planning Commission (1956) Second five year plan. Planning Commission, India

Planning Commission (1961) Third five year plan. Summary

Planning Commission (1970) The fourth five-year plan 1970–75

Planning Commission (1979) Report of the working group on energy policy. Planning Commission, Government of India, New Delhi

Planning Commission (2001) Indian planning experience—a statistical profile

Planning Commission (2015) 12th five year plan (2012–17)

Rajya Sabha (2016) Starred question no. 897. Govt. of India

Sinha SK, Subramanian KA, Singh HM, Tyagi VV, Mishra A (2019) Progressive trends in bio-fuel policies in India: targets and implementation strategy. Biofuels 10(1):155–166

Srivastav A (2019) The science and impact of climate change. Springer

US EIA (2018) International energy outlook 2018 (IEO 2018) Key takeaways. US Energy Inf. Adm. http://www.eia.gov/pressroom/presentations/capuano_07242018.pdf

Watanabe J, Bloomberg New Energy Finance (2016) Giant fall in generation costs from offshore wind. Bloomberg new energy finance 2:2

WEO, IEA (2006) World Energy Outlook 2006. International Energy Agency, IEA-OECD

World Bank (2013) Little green data book 2013

World Bank (2017) Little green data book 2017

World Economic Forum (2017) Global energy architecture performance index report 2017. WEF and Accenture

World Nuclear Association (2014) World Nuclear Association Website

Chapter 4
Climate Mitigation and India's Commitment to Global Community

Abstract The 1972 Stockholm declaration on the human environment led to constitution of first-generation conventions such as CITES, CMS, and WHC followed by the second-generation conventions such as CBD and UNFCCC that were adopted post 1991 as an outcome of Earth Summit and led to the development of holistic approaches to conservation reflecting the sensitivities of poverty, rural development, equal participation, sustainable use, education, and awareness. The UNFCCC makes it obligatory on all signatories to develop, publish, periodically update, and make available national inventory of GHGs, steps taken or envisaged to implement the Convention. India submitted its Initial National Communication (INC) to the UNFCCC in UNFCCC website https://unfccc.int entailing emissions inventory of 1994. Second National Communication (SNC) was submitted to the UNFCCC in 2012 wherein GHG inventory for the year 2000 was reported and a summary of GHG inventory for the year 2007 was provided as a proactive measure. Thereafter, India submitted its first and second biennial updated reports to the UNFCCC in 2015 and 2018, respectively, in accordance with the decision taken in the 16th session of the CoP to the UNFCCC.

Keywords Earth summit · Soft laws · UNFCCC · India submissions · National action plan · Climate vulnerability

4.1 Multilateral Environmental Affairs

Early 1970s was the period when the world was heavily polarized and the global conservation priorities were mainly focused on the issues of wildlife management, land degradation, soil conservation, and water pollution. Developing and underdeveloped countries viewed environmental concerns as western responsibility and the communist bloc relentlessly continued destruction of environment. Under these circumstances the global assembly in Stockholm in 1972 was a surprising breakthrough and the outcome was known as the 'Stockholm spirit of compromise' in which developed and developing world found ways of resolving divergent and

© Springer Nature Singapore Pte Ltd. 2021
A. Srivastav, *Energy Dynamics and Climate Mitigation*,
Advances in Geographical and Environmental Sciences,
https://doi.org/10.1007/978-981-15-8940-9_4

conflicting issues for the sake of posterity. The Swedish Prime Minister Olof Palme exhorted the delegates by his following remarks:

People are no longer satisfied with declarations. They demand firm actions and concrete results. They expect that the nations of the world, having identified a problem, will have the vitality to act.

The above comments were possibly based on the hard work and collection of scientific evidence by many scientists since mid-nineteenth century as described in Box 1.

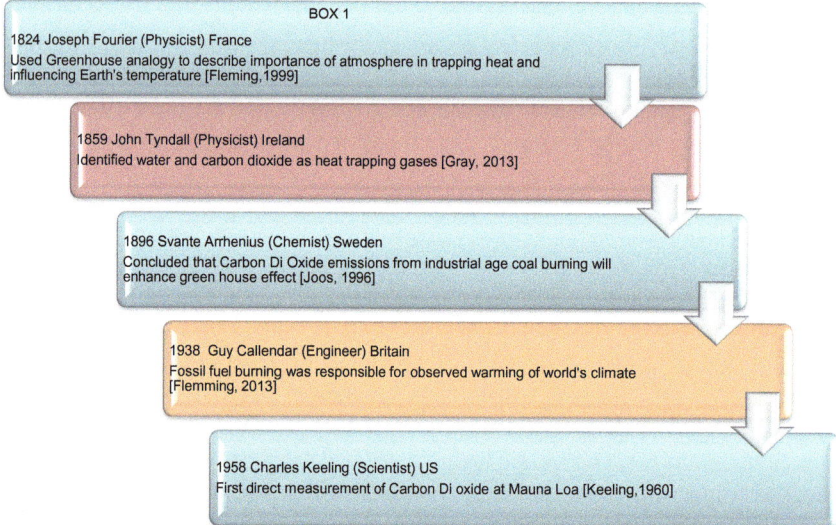

BOX 1

1824 Joseph Fourier (Physicist) France
Used Greenhouse analogy to describe importance of atmosphere in trapping heat and influencing Earth's temperature [Fleming,1999]

1859 John Tyndall (Physicist) Ireland
Identified water and carbon dioxide as heat trapping gases [Gray, 2013]

1896 Svante Arrhenius (Chemist) Sweden
Concluded that Carbon Di Oxide emissions from industrial age coal burning will enhance green house effect [Joos, 1996]

1938 Guy Callendar (Engineer) Britain
Fossil fuel burning was responsible for observed warming of world's climate [Flemming, 2013]

1958 Charles Keeling (Scientist) US
First direct measurement of Carbon Di oxide at Mauna Loa [Keeling,1960]

Environment and conservation acquired respect and significance after Stockholm conference and almost 110 countries set up environment department and ministries within 10 years. The Stockholm declaration on the human environment constituted the first body of 'soft law' in international environmental affairs and the conventions (such as CITES, CMS, and WHC) that were adopted post 1972 Stockholm conference are known as first-generation conventions. Unfortunately, several events of 1970 and 1980s overshadowed the achievements of 1972 as the sub-Saharan region fell behind in improving per capita income due to a combination of factors including severe droughts and unfavorable trade. Almost one million humans starved to death in Ethiopia during 1984–1985 drought and many biodiversity-rich countries in Africa, Asia, and Latin America registered high population and low economic growth, suffered from wars, internal conflicts, natural disasters, and environmental pollution. The environmental catastrophe caused by 1991 Gulf War was a blow to the conservation world as more than 30,000 sea birds perished in oil slick that also contaminated large areas of mangroves and coral reefs. The environmental agenda almost collapsed during this period as deforestation and degradation continued, many species lost their habitats, pollutants damaged land, air, and water with impunity.

Nonetheless, this period brought together the conservation community and strengthened the global view that biodiversity conservation and environmental pollution control required long-term, collaborative, and holistic actions. The Stockholm milestones were resuscitated after the publication of Brundtland Commission report titled 'Our Common Future' (Brundtland 1987) followed by 1992 the Earth Summit in Rio de Janeiro (Summit 1992). The Earth Summit led to the development of second-generation conventions such as CBD and UNFCCC that gave stress on the holistic approaches to conservation reflecting the sensitivities of poverty, rural development, equal participation, sustainable use, education, and awareness. The Brundtland Commission's report essentially exposed the myth that industrial development and environmental conservation were complimentary. It also warned that natural tropical forests had been ruthlessly destroyed at an annual rate of 11 million hectares during last three decades, the use of fossil fuels grew nearly 30-fold, and industrial growth increased more than 50-fold. There was hardly any realization that the cost of repair and restoration from environmental damage was extremely high and in some cases the process was irreversible as well. The burning of fossil fuels and forest biomass was eventually responsible for global warming leading to sea level rise over the next 45 years threatening to inundate low-lying countries and disrupt national agriculture production and trade.

Ten years after Stockholm Declaration, the UN General Assembly, on October 28, 1982, adopted the World Charter for Nature (Assembly 1982), to respect every form of life and the uninterrupted functioning of natural systems on which life depends and thus called for maintaining the stability and quality of nature and natural resources. The Charter cautioned that degradation of natural systems through overconsumption and misuse of natural resources as well as failure to establish appropriate economic order may lead to breakdown of social, economic, and political framework of civilization. During this period of industrialization and environmental destruction, all such nations that had protected their environment at the cost of economic development formed a close-knit group and divided the world in two parts—the South (the third world) and the North (the industrialized world). The third world countries often blamed the rich industrial nations for deforestation and pollution, whereas the industrial nations blamed the third world for lack of necessities, infrastructure, energy, scientific and technological advancement, and so on. Eventually, the desire for survival, opulence, and replication of human genome forced the conservation of nature and natural resources to take a back seat in the interest of economic development.

After almost five decades, the humanity is at crossroads facing serious challenges as the earth heats up.

Forests and its biological as well as non-biological resources have played a pivotal role in building global economy prior to the advent of industrial revolution. This role changed after the discovery of coal and oil and its commercial use during the industrial revolution. Industrial expansion led to massive destruction of natural aquatic and terrestrial ecosystems, especially during the First and Second World Wars. It was only after the global conservation experts expressed serious concerns at the rapid disappearance of flora and fauna and threat of annihilation due to global warming, that several platforms for brainstorming and to evolve ways for their protection were established and effective measures were taken by many countries to conserve their biological resources. Eventually, last three decades have witnessed a series of legal and non-legislative measures adopted at the global, regional, as well as national levels.

Twenty years after Stockholm and ten years after the World Charter for Nature, the United Nations convened a conference in Rio de Janeiro, Brazil in June 1992 to reiterate the Stockholm declaration as well as find ways of integrating economic development and environmental protection. Two conventions were drafted at Rio: the UN Convention on Climate Change (UNFCCC) and the Convention on Biological diversity (CBD). The UNFCCC (United Nations Framework Convention on Climate Change), as the title indicates, is a framework convention which provides for the basic obligations but leaves the specific commitments to the signatories. The Convention follows a strange concept of 'common but differentiated responsibilities' for stabilizing greenhouse gas concentration in the atmosphere at a level that would prevent dangerous anthropogenic interference with the natural system. One of the important aspects of this Convention is the premise that ecosystems may adapt naturally to climate change and therefore *prevention of dangerous anthropogenic interference with the climate system* does not imply *prevention of climate change*. The principles mentioned in Article 3 of the Convention are in the form of recommendation ('should') and provides scope for interpretations/misinterpretations. These are:

i.   The Parties should protect the climate system for the benefit of the present and future generations of humankind on the basis of equity and in accordance with the common but differentiated responsibilities and respective capabilities.
ii. The specific need and special circumstances of developing country Parties should be given full consideration.
iii. The Parties should take precautionary measures to anticipate, prevent, or minimize the causes of climate change and mitigate its adverse effects.
iv. The Parties have a right to, and should, promote sustainable development.
v.   The Parties should cooperate to promote a supportive and open international economic system that would lead to sustainable economic growth and development in all Parties particularly developing country Parties.

The UNFCCC established three different groups of nations and categorized them in three annexures:

- Annex 1 nations—consist of OECD member nations and economies in transition.
- Annex 2 nations—consist of only OECD member nations that are required to provide financial resources to developing countries for emission reduction.
- Annex 3 nations—consist of mostly developing countries.

Each group of nations has been given certain responsibilities that are mentioned in Box 2 [Ref—UNFCCC website https://unfccc.int].

Box 2 [Source-https://unfccc.int]

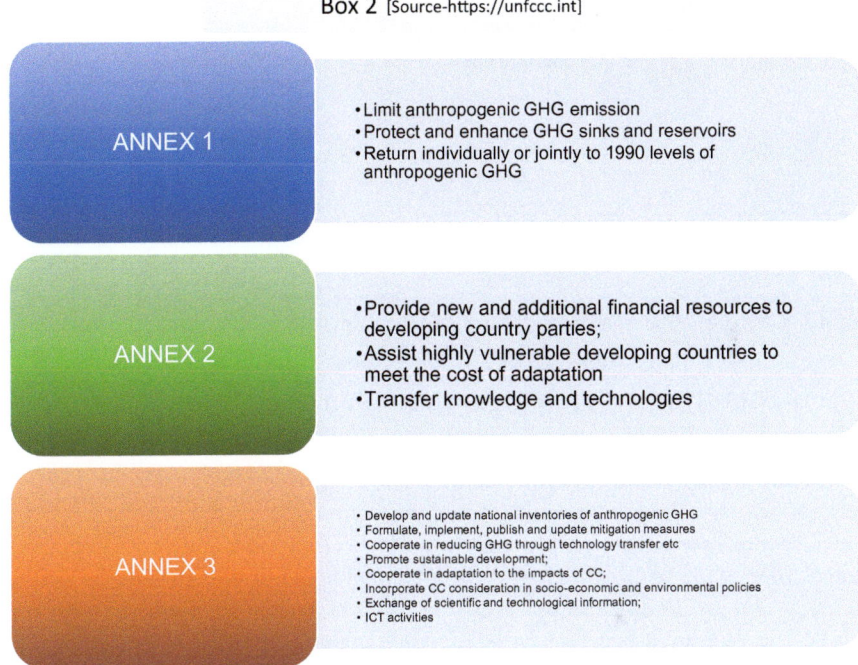

ANNEX 1
- Limit anthropogenic GHG emission
- Protect and enhance GHG sinks and reservoirs
- Return individually or jointly to 1990 levels of anthropogenic GHG

ANNEX 2
- Provide new and additional financial resources to developing country parties;
- Assist highly vulnerable developing countries to meet the cost of adaptation
- Transfer knowledge and technologies

ANNEX 3
- Develop and update national inventories of anthropogenic GHG
- Formulate, implement, publish and update mitigation measures
- Cooperate in reducing GHG through technology transfer etc
- Promote sustainable development;
- Cooperate in adaptation to the impacts of CC;
- Incorporate CC consideration in socio-economic and environmental policies
- Exchange of scientific and technological information;
- ICT activities

The ultimate objective of the United Nations Framework Convention on Climate Change (UNFCCC) is to achieve, in accordance with the relevant provisions of the Convention, stabilization of greenhouse gas concentrations in the atmosphere at a level that would prevent dangerous anthropogenic interference with the climate system. Reduction in GHG emission is expected to be achieved within an agreed timeframe so as to allow ecosystems to adapt naturally to climate change, to ensure that food production is not threatened, and to enable economic development to proceed in a sustainable manner [https://unfccc.int].

The UNFCCC makes it obligatory on all signatories to develop, publish, periodically update and make available national inventory of GHGs (not controlled by the Montreal Protocol), steps taken or envisaged to implement the Convention and any other information that the Party considers relevant to the achievement of the objective of the Convention and suitable for inclusion in its communication. Subsequently, it was also decided to enhance reporting from Parties not included in Annex I to the Convention. Paragraph 60(c) of this decision stated that 'developing countries, consistent with their capabilities and the level of support provided for reporting, should submit biennial update reports containing updates of national greenhouse gas inventories, including a national inventory report and information on mitigation actions, needs and support received'. The decision also stated 'the need to consider national capability and financial support required to facilitate the timely preparation of biennial update reports'.

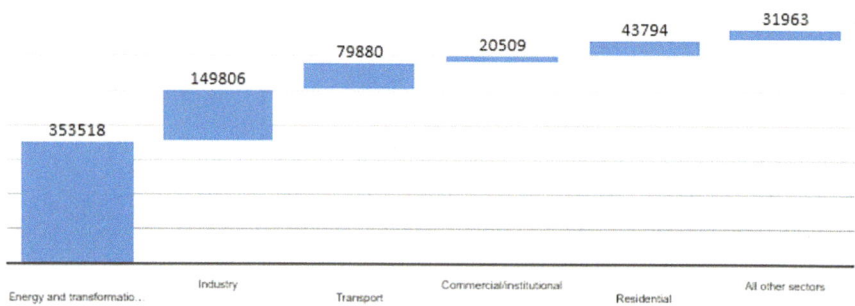

Fig. 4.1 Carbon di oxide emission from energy sector (Giga gram). (*Source*—India, MoE 2004)

4.2   India's Initial Submission to UNFCCC in 2004

India made its first submission[1] in the year 2004 specifying inventory for the base year 1994 as stipulated in the guidelines of the Convention. During that period India had the world's second largest population and fourth largest economy with per capita annual GDP of US $462. Three quarters of its population lived in rural areas having vast informal and traditional sectors and weak market that coexisted with growing formal and modern urban sector. India's population surpassed one billion by the turn of twenty-first century with largely agrarian mass with poor literacy rate. The compulsion of economic development demanded that the nation's energy use during the second half of the twentieth century expanded with a shift from non-commercial to commercial energy. Sectors such as transport, construction, road, irrigation, steel and fertilizer demanded intensive energy most of which was based on use of fossil fuels (mainly coal and petroleum products) that contributed 95% of the total commercial energy consumed in India in 1994. The remaining 5% was derived from sources like hydropower, nuclear, and renewable energy. Use of fossil fuels contributed to 91% carbon dioxide emission with 62% contributed by coal, 31% by petroleum, and 7% by natural gas.[2]

The graphic details of carbon dioxide status from various sources are provided in Figs. 4.1, 4.2, and 4.3.

India, being a low-income economy, depended heavily on coal for energy generation for two reasons: indigenous availability and production and low cost. With significant coal reserves of around 221 billion tons, the share of coal in generating

---

[1]India submitted its Initial National Communication (INC) to the UNFCCC in 2004 entailing emissions inventory of 1994. Second National Communication (SNC) was submitted to the UNFCCC in 2012 wherein GHG inventory for the year 2000 was reported. In addition, a summary of GHG inventory of the year 2007 was provided as a proactive measure.

[2](*Source* India, MoE 2004).

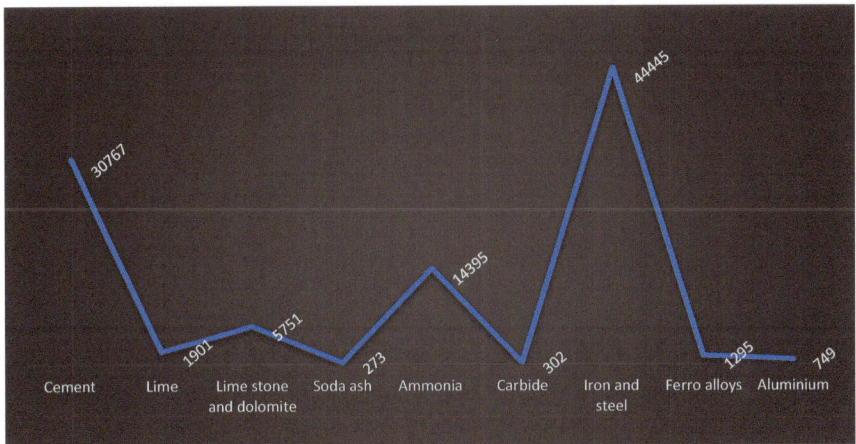

Fig. 4.2 Carbon di oxide emission from Industrial processes (Giga gram). (*Source*—India, MoE 2004)

Fig. 4.3 Carbon di oxide emission and removal (Giga gram) from land use, land use change and forestry. (*Source*—India, MoE 2004)

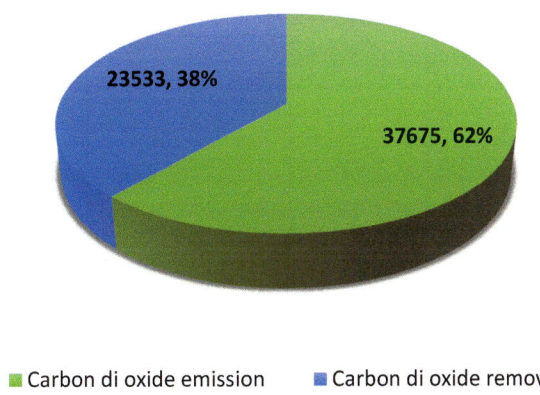

■ Carbon di oxide emission   ■ Carbon di oxide removal

electricity was 47% followed by petroleum (20%) and natural gas (11%). Nearly 70% of the power requirement in India was then supplied by coal-based thermal power plants. Figure 4.4 gives an overview of consumption of various fossil fuels over four decades.

Petroleum invaded Indian commercial market sometime after 1950 and gained prominence rapidly. From a share of mere 2% in 1953–1954 (100% imported into India at that time), it went up to about 27% in 2001–2002. The share of natural gas also increased from almost 0 to 6% during the same period. While petroleum essentially fed Indian transport sector, electricity required for development and household use was generated from a diversity of energy mix with an increase in capacity by almost seven-fold between 1970 and 2000. The capacity mix in year 2000 included 61% coal, 8% gas, 2% nuclear, 1.5% renewables, and 24% hydro-based power. Other than consumption of fossil fuel energy, about 90% of the rural and 30% of urban

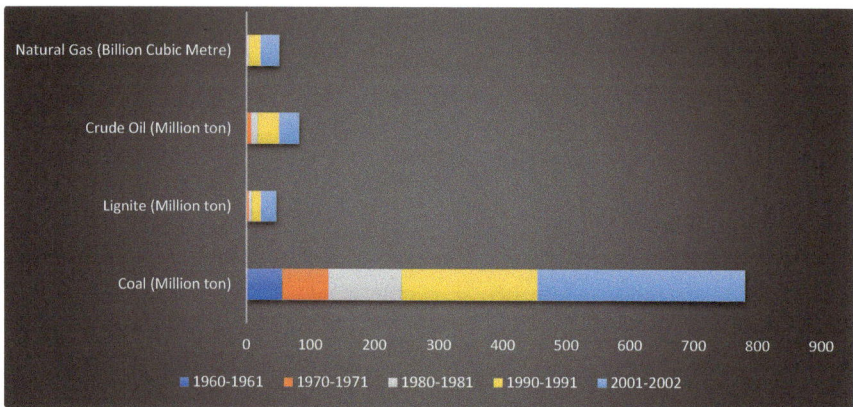

Fig. 4.4 Trends in commercial energy sources. (*Source*—India, MoE 2004)

households in India consumed large quantity of traditional fuels or non-commercial energy or non-conventional fuels such as firewood, dung cake, and agriculture waste. Between 1953 and 2001, the total primary energy supply grew at 3.4% annually reaching 437.7 million tons of oil equivalent (Mtoe) in 2001.

The initial submission admits the fact that use of coal will continue to be the mainstay of commercial energy in India. Coal in India is classified into three main categories—coking, non-coking, and lignite with total reserve of about 206 billion tons (up to a depth of 1200 meters). However, the recoverable coal reserves are estimated to be 75 billion tons, which is enough for supplying coal for over 250 years at current levels of production. Lignite reserves, on the other hand, have been estimated at around 34,763 million tons of which 30,275 million tons are recoverable. Two significant factors have contributed to the rising carbon emissions in India. One is the large proportion of coal in energy (mainly power generation) production and the other is sharp rise in the demand for power. Nonetheless, India's per capita CO_2 emission of 0.87 t-CO_2 in 1994 was among the low-emission nations.[3] India's energy, power, and carbon intensities of the GDP have declined after the mid-1990s, due to factors such as increased share of service sector in the GDP, and energy efficiency improvements.

India emitted 1,228,540 Gg (gigagram) of greenhouse gases in 1994 with largest contribution (61%) by energy sector, followed by agriculture (28%), industrial process (8%), waste (2%) and land use, land use change, and forestry (1%). Following are the highlights of aggregate GHG emissions from the anthropogenic activities in India:

1. *Carbon dioxide = 793,490 Gigagram*
2. *Methane = 18,583 Gigagram*
3. *N_2O = 178 Gigagram.*

---

[3]India's per capita share of CO_2 emission in 1994 was 4% of the USA, 8% of Germany, 9% of UK, 10% of Japan, and 23% of the global average.

On a sectoral basis.[4]

1. *Energy sector = 743,820 Gigagram of CO_2 equivalent (61% of total);*
2. *Agriculture sector contribution = 3,44,485 Gigagram of CO_2 equivalent (28% of total);*
3. *Industrial sector contribution = 1,02,710 Gigagram of CO_2 equivalent (8% of total);*
4. *Waste disposal contribution = 23,233 Gigagram of CO_2 equivalent (2% of total);*
5. *Land use, land use change, and forestry sector contribution = 14,292 Gigagram of CO_2 equivalent (1% of total).*

The details of CO_2 emissions in 1994 by various sectors of economic development are as follows:

i.   Industrial sector: Paper, sugar, cement, iron and steel, textile, bricks, fertilizer, chemical, aluminum, ferroalloys, non-ferrous, food and beverages, leather and tannery, jute, plastic, mining and quarrying, rubber, and others use coal and petroleum products as energy sources in substantial quantities. The total CO_2 emission from this sector was 149,806 Gg.

ii. Commercial sector: Activities like cooking, lighting, space heating/cooling, refrigeration, and pumping characterize the commercial sector that use electricity, LPG, kerosene, diesel, coal, charcoal, and fuelwood. The total CO_2 emission from this sector was 20,509 Gg.

iii. Residential sector: The residential sector consumes energy for lighting, cooking, heating/cooling, and household appliances. In Indian context the energy source ranges from biomass-based fuels to coal, kerosene, gas, and electricity. The total CO_2 emission from this sector was 43,794 Gg. The emission from biomass burning was excluded for the strange reason of carbon neutrality.

iv. Transport sector: Road transport contributed almost 90% of the total emission of transport sector which was 79,880 gigagram in 1994. Large number of petrol and diesel vehicles spew toxic pollutants as well as CO_2, CH_4, and N_2O. Corrective measures were taken subsequently such as introduction of CNG vehicles, improvement in auto fuel quality, and enhancement of road infrastructure.

v.   Other sectors: The emission from sectors not included anywhere above was 31,963 gigagram.

vi. Industrial processes:

   a. The major industrial processes responsible for CO_2 emission (Fig. 4.5) included cement, iron and steel, lime, limestone and dolomite use, soda ash manufacture, ammonia, ferroalloys, aluminum and manganese foundries, and calcium carbide production. The total CO_2 emissions from the industrial processes were estimated to be 99,878 gigagram in 1994. Cement, iron, and steel manufacturing processes were the key source for CO_2 emissions (75,212 gigagram CO_2).

---

[4]CO_2 emissions from biomass fuels were treated as carbon neutral in the report and therefore not included in the national totals.

Process	Emission (Giga gram) in 1994
Cement	30767
Lime	1901
Limestone and Dolomite use	5751
Soda ash use	273
Ammonia production	14395
Carbide production	302
Aluminum production	749

Fig. 4.5 Industrial processes responsible for CO_2 emission. (*Source*—India, MoE 2004)

4.3 Land Use, Land-Use Change, and Forestry

The IPCC[5] approach to assess changes in carbon stock encompasses four aspects, viz.:

- Changes in forest and other woody biomass stocks.
- Conversion of forest and grassland.
- Uptake from abandonment of managed lands.
- Emissions and removals from soils.

The area under forests (including tree plantations) in India was estimated to be 63.33 million hectares in 1994 and the annual loss of biomass due to conversion of forest to agriculture land or for developmental purposes and so on was estimated to be 12.09 terragram. Accordingly, the total quantity of carbon dioxide released from conversion was estimated to be 17,987 Gg. The total soil carbon stock has been estimated for all land-use systems in India for the period between 1984 and 1994 and the difference is considered as net emission of CO_2 for 1994. Following this methodology, the net change in soil carbon stock in mineral soils averaged over the decade (1984–1994) has been estimated to be 19.68 Tg CO_2 (India, MoE 2004).

The details of methane CH_4 emissions in 1994 by various sectors of economic development are given in Box 3.

---

[5]IPCC – or Intergovernmental Panel on Climate Change is the United Nations body established to assess the science and related aspects of climate change.

BOX 3 - Details of Methane CH₄ emissions in 1994	
[Source- India, MoE. 2004]	
Rice cultivation	Anaerobic decomposition of organic material in flooded rice fields produces methane, which escapes into the atmosphere primarily by diffusive transport through the rice plants during the growing season. There are large spatial and temporal variations of methane fluxes which occur due to different soil types, soil organic carbon and various agricultural practices such as choice of water management and cultivar, the application of organic amendments, the mineral fertilizer, and soil organic carbon.
Burning of Agricultural crop residue	Burning of crop residue is a significant net source of methane. The amount of agricultural waste produced by a country depends on its crop management system. In India, the primary end-uses of crop residue are as animal fodder, industrial and domestic fuel, thatching, packaging, bedding, construction of walls/fences, and as green-manure and compost. The leftover is available for field burning and even this varies with local and regional climate, season, livestock distribution, availability of fuelwood, availability of fodder, weed infestation etc.
Manure management	The decomposition of organic animal waste in an anaerobic environment produces methane. The waste produced by non-ruminant animals in India is not collected. However, that of cattle and buffalo are used for a variety of purposes including sun dried dung cakes for cooking or for biogas.
Municipal Solid waste management	Solid waste disposal in India takes place in two distinct ways. In rural areas and small towns, there is no systematic collection of waste and it is haphazard. As anaerobic conditions do not develop, no methane is generated in these areas. However, in urban towns, solid waste is disposed by land filling in low-lying areas located in and around the urban centers. Due to stacking of waste over the years, anaerobic conditions develop, and hence these dumping sites generate large quantities of biogas containing a sizeable proportion of methane.

There were several sources contributing to methane release (Fig. 4.6) in the atmosphere, including exhaustive mining of coal, emissions due to oil and natural gas, products derived from livestock, increased production of rice to meet the demand of the growing population, on-site burning of crop residue, management of solid waste and waste water from the domestic and the industrial sectors. Methane emission during the year 1994 was 18,583 gigagram and most of it (78%) was released by agriculture sector (primarily enteric fermentation and rice cultivation. Other sectors of methane emission included biomass burning, coal mining, and flaring of natural gas were responsible for 16% and waste disposal activities contributed to about 6%. Methane emitted from land use, land-use change, and forestry sector (due to the

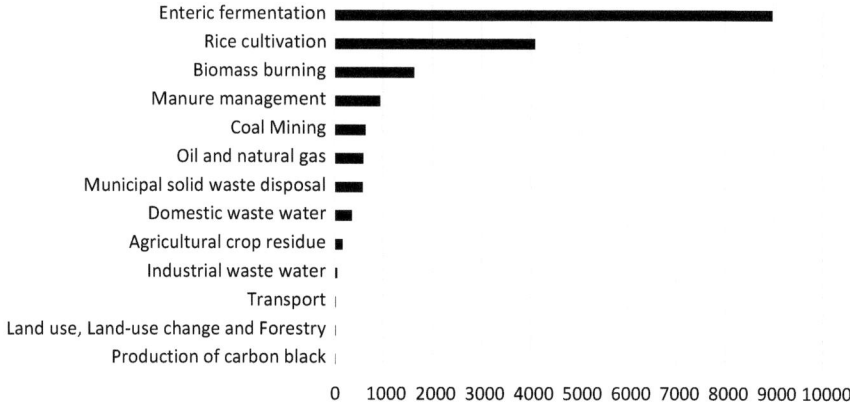

Fig. 4.6 Methane emission (Gigagram). (*Source*—India, MoE 2004)

burning of biomass in shifting cultivation) as well as from industrial processes was insignificant.

The details of N_2O emission in 1994 are as follows:

N_2O emission in 1994 was 178 gigagram contributing 4% of the total GHG emissions. Agriculture sector was the major contributor accounting for 84%, followed by fuel combustion (7%), industrial processes (5%), and waste (4%).

4.4   First Biennial Report[6] by India—2015

India submitted its first biennial updated report to the UNFCCC in 2015 in accordance with the decision taken in the 16th session of the CoP to the UNFCCC. The report contains the following major components with GHG inventory for the year 2010:

(a) National circumstances.
(b) National greenhouse gas inventory.
(c) Mitigation actions.
(d) Domestic monitoring, reporting, and verification (MRV) arrangement and
(e) Finance, technology, and capacity building needs and support received.

India's net GHG emission in 2010 was 2,136,841.24 gigagram of CO_2 equivalent (2,136.8 million tons of CO_2 eq.) from the energy, industrial processes and product use, agriculture, and waste sectors (excluding LULUCF). The largest share of GHG emission was contributed by the energy sector (70.7%), followed by the agriculture (18.3%), industrial process (8%), and waste (3%). Land use, land-use change, and forestry sectors were a net sink in 2010. Following are the highlights aggregate emissions from the anthropogenic activities:

---

[6](*Source*: MoEFCC 2015).

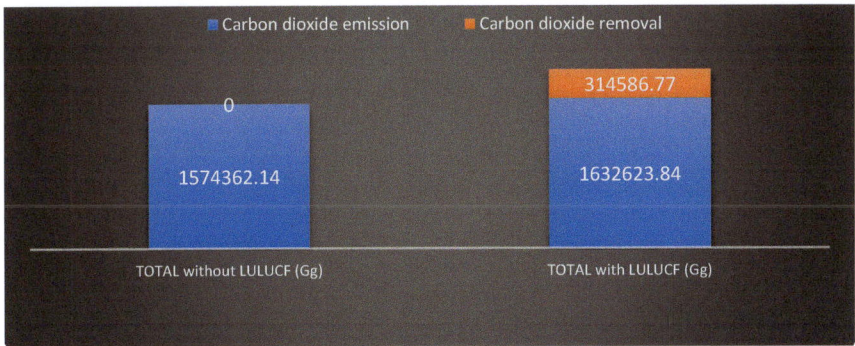

Fig. 4.7 Carbon dioxide emission and removal (Gigagram) in 2010. (*Source*—MoEFCC 2015)

	CO_2 emission	CO_2 removal	Methane	Nitrogen Oxides	HFC 23	CF4	C2F6	SF6
TOTAL without LULUCF (Gg)	1574362	0	19623	368	1.43	2.13	0.58	0.0042
TOTAL with LULUCF (Gg)	1632623	314586	19776	370	1.43	2.13	0.58	0.0042
Energy	1441882	0	2534	48	0	0	0	0
Fuel Combustion Activities	1441882	0	204	48	0	0	0	0
Energy Industries	876180	0	11	12	0	0	0	0

Fig. 4.8 National greenhouse gas inventory of anthropogenic emissions by sources and removals by sinks (Gigagrams). (*Source*—MoEFCC 2015)

1. Carbon dioxide = 1,574,362 gigagram (Fig. 4.7)
2. Methane = 19,623 gigagram
3. N_2O = 368 gigagram.

The details of emission from different sectors[7] are as follows:

1. Energy sector = 1,510,120 gigagram of CO_2 equivalent of GHGs (70.7% of total).
2. Agriculture sector = 390,165 gigagram of CO_2 equivalent of GHGs (18.3% of total).
3. Industrial sector = 171,502 gigagram of CO_2 equivalent of GHGs (8% of total).
4. Waste disposal = 65,052 gigagram of CO_2 equivalent of GHGs (3% of total).
5. Land use, land-use change, and forestry sector were a net sink.

A summary of emissions from these sectors is presented in Fig. 4.8.

---

[7]CO_2 emissions from biomass fuels were treated as carbon neutral in the report and therefore not included in the national totals.

The report mentions that the total installed capacity for power generation in March 2015 was 271,722 megawatts[8] with coal as the primary energy source for power generation. It reiterates the fact that coal will continue to fuel thermal power plants in India in future and in order to improve energy efficiency of the coal-based power plants and to reduce the GHG emissions. The new thermal power plants would be based on super-critical technology. As far as transport sector is concerned, India largely depended on import of petroleum that increased from 74.10 million metric tons in 2000–2001 to 189.44 million metric tons in 2014–2015.

4.5   Commitment

In accordance with the provisions of Article 12, paragraphs 1(b) and 4, and Article 10, paragraph 2(a), of the Convention, India communicated its voluntary pledge to reduce the emissions intensity of its GDP by 20–25% by 2020 compared with the 2005 level. The emission intensity of GDP was reduced from 35.14 kilogram of CO_2 eq./1000 in 2005 to 31.01 kilogram of CO_2 eq./1000 in 2010. The voluntary commitment to reduce emission intensity excluded emissions from agriculture. A national action plan for climate change was launched in 2008 to mitigate climate impact (Government of India. National Action Plan 2008). Summary of the plan is provided in Fig. 4.9.

A summary of actions/proposals to mitigate climate action is provided:

i   Development of ultra-mega power projects of about 4,000 megawatts capacity each using super-critical technology with the intention of achieving higher levels of efficiency, substantial fuel saving and reducing GHG emission. Also, 40 super-critical plants with a total capacity of 27,485 megawatts have been installed.

ii   Introduction of super-efficient equipment programme (SEEP) through efficient fans, LED bulbs, and tube lights. This was expected to save 6.06 billion units per year by 2016–2017 and help offset an installed capacity of 1500 megawatts during 12th five-year plan.

iii The overall target for renewable energy was enhanced from 35,775 megawatts in 2015 to 175,000 by 2022 comprising 100,000 megawatts solar, 60,000 megawatts wind, 10,000 megawatts biomass, and 5000 megawatts small hydropower, respectively.

iv   During the year 2014–2015, India generated 37,835 million units of energy from 21 nuclear power plants with installed capacity of 5780 megawatts. This capacity was expected to be enhanced to nearly 10,000 megawatts by 2018–2019 and thereafter to 63,000 megawatts by 2032.

v   Five million incandescent lamps were replaced by compact fluorescent lamps (CFL) in 2011 saving 231 million units of electricity. It was intended to replace

---

[8]In 2010, the installed capacity of renewable energy-based grid-interactive power-generating units was 35,777 megawatts.

Name of the Mission	Key objectives	Targets
Jawaharlal Nehru National Solar Mission	Establish India as a global leader in solar energy by creating the policy conditions for its diffusion across the country as quickly as possible.	GW energy from solar by 2022. 20 million solar lighting systems for rural areas by 2022
National Mission on Sustainable Habitat	• Energy efficiency in residential and commercial sectors. • Switch over to public transport by commuters. • Recycling of material and urban waste management: power from waste	• Increasing energy efficiency in buildings: building bye laws and standards, energy performance monitoring, national standards for construction and recycling of construction waste. • Urban transport: norms integrating congestion charges, parking, norms for pedestrian and cycling, integrating transport planning with spatial planning. • Water supply: mandatory rainwater harvesting, water and energy audits.
National Mission on Green India	• Increase forest/tree cover by 5 million hectares. • Improve quality of forest/tree cover on another 5 million hectares. • Improve eco-system services like carbon sequestration. • Increase forest-based livelihood income of about 3 million households. • Enhanced annual CO_2 sequestration by 50-60 million tonnes in the year 2020	• Qualitative improvement of forest cover/ecosystem in moderately dense forests (1.5 Mha), open degraded forests (3 Mha), degraded grassland (0.4 Mha) and wetlands (0.1 Mha). • Eco-restoration/afforestation of scrub, shifting cultivation areas, cold deserts, mangroves, ravines and abandoned mining areas (1.8 Mha). • Bringing urban/peri-urban lands under forest and tree cover (0.20 Mha). • Agro-forestry/social forestry (3 Mha). • Improvement of forest-based livelihoods for about 3 million households living in and around forests.
National Mission on Sustaining Himalayan Ecosystem	Develop a sustainable national capacity to continuously assess the health status of the Himalayan Ecosystem.	• Continuous monitoring of the eco-system. • Data generation. • Glaciology research • Generation of bio-geo database • Ecological modeling of Himalayas. • Prediction of Socio-Economic and Climate change scenarios. • Vulnerability assessment.
National Mission on Enhanced Energy Efficiency	Upscale the effort to unlock the market for energy efficiency. Cumulative avoided electricity capacity addition of 19,598 MW	• Specific energy consumption reduction. • Incentivizing action through energy savings certificates. • National energy efficiency road.

Fig. 4.9 National action plan. (*Source*—Government of India. National Action Plan 2008)

400 million incandescent lamps including 30 million streetlamps in the country
and save 18,400 million units per year.

vi The telecom regulatory authority of India pledged to use renewable energy
technologies and grid power for at least 50% rural and 20% urban towers by
2015 to be scaled up to 75% rural and 33% urban towers by 2020. In addition,
the authority asked the service providers for adopting energy-efficient network
planning, adoption of renewable energy, and development of energy-efficient
technologies.

vii An auto fuel policy was introduced in 2003 to address the energy and environmental challenges in the automobile industry. Accordingly, fuel efficiency
norms have been introduced from time to time to address energy and environment concerns. A national electric mobility mission plan 2020 has been
developed to increase the share of CNG vehicles to 30–35% by 2020 as well
as to increase electric vehicles.

viii An initiative on climate resilient agriculture (NICRA) was adopted in 2011 to
make farmers adapt to climate change, improve crop production, reduce GHG
emission, and sustainable management of natural resources.

ix Substantial reduction in diversion of forest land (from 4.13 million hectare
between 1951 and 1976 to 1.133 million hectares between 1980 and 2011) has
improved carbon content in forests trees and soil.

x Improve one million hectare each of open (less than 10% crown canopy) and
medium dense (between 10 and 40% crown canopy) to next higher category.

xi Sixty cities to be developed as solar/green cities.

xii India made a voluntary pledge in 2010 to reduce the emission intensity of
its GDP by 20–25% from 2005 levels by 2020 (excluding emissions from
agriculture).

4.6 Second Biennial Report by India—2018

The second biennial updated report was submitted by India to the UNFCCC in 2018
(MoEFCC 2018). The report contains the following major components with GHG
inventory for the year 2010:

(f) National circumstances
(g) National greenhouse gas inventory
(h) Mitigation actions
(i) Domestic monitoring, reporting and verification (MRV) arrangement
(j) Finance, technology, and capacity building needs and support received
(k) Additional information.

As compared with the first biennial report where inventory for the year 2010 was
provided, the second report cover data till 2014. India's net GHG emission registered
marginal decline from 2,136,841.24 gigagram of CO_2 equivalent (2,136.8 million
tons of CO_2 eq.) in 2010 to 2,607,488 Gg from the energy, industrial processes and

	CO₂ emission	CO₂ removal	Methane	Nitrogen Oxides	HFC 23	CF4	C2F6	SF6
TOTAL without LULUCF (Gg)	1997891.85	-	20005.35	475.29	1.59	2.61	0.71	0.004
TOTAL with LULUCF (Gg)	2015107.88	319860.23	20053.54	476.71	1.59	2.61	0.71	0.004
Energy	1844705.03	-	2133.37	65.35	-	-	-	-
Industrial Processes and Product Use	153186.81	-	177.85	10.36	1.59	2.61	0.71	0.004
Agriculture	-	-	14709.78	349.39	-	-	-	-
LULUCF	17216.04	319860.23	48.19	1.42	-	-	-	-
Carbon di oxide from Biomass	807087.06	-	-	-	-	-	-	-

Fig. 4.10 National greenhouse gas inventory of anthropogenic emissions by sources and removals by sinks (Gigagrams). (MoEFCC 2018)

product use, agriculture, and waste sectors (excluding LULUCF) in 2014. The energy sector accounted for 73% of the total GHG emissions and fuel combustion activities emitted 1,871,709 Gg CO_2 equivalent in 2014 including 1,140,983 CO_2 equivalent from energy industries. Within energy industries, 94.96% of emissions were from electricity production, 4.39% from refinery, and 0.66% from manufacturing of solid fuels. Electricity production accounted for almost 42% of entire GHG emissions. A summary of emissions from different sectors is given in Fig. 4.10.

India implemented its mitigation measures mostly through policy and legal instruments, technology development, standards to reduce vehicular emissions, and the strategy to increase the share of alternative fuels in the overall fuel mix. India reported 11 mitigation actions for the energy sector mainly in the areas of renewable energy and energy efficiency in the power sector, as the power sector contributes 43% of India's total GHG emissions. Of the mitigation actions that were quantified, the highest GHG emission reduction was achieved from the national program for LED-based home and street lighting that replaced 312 million incandescent lamps for street lighting and reduced 33 Mt CO_2 eq. per year. Mitigation action in the transport sector, which is one of the fastest growing sectors in India, has substantially reduced GHG emission. For example, metro rail and rapid transit system helped in achieving emission reduction of 0.72 and 0.044 Mt CO_2 eq. in 2014. Nine mitigation actions in the forestry sector, mostly plantations, afforestation, reforestation, and programs to conserve and enhance the carbon sink, were driven by various legal instruments. Under this sector, India also reported 19 CDM projects that were registered under afforestation or reforestation with annual emission reduction potential was 429,614 tons of CO_2 eq. Similarly, National Horticulture Mission had a major impact and led to the sequestration of 137 Mt CO_2 from 2010 to 2016.

In 2018, India developed a national REDD + strategy with the overarching objective of facilitating the implementation of REDD + under the Convention. The REDD + strategy was aligned with the national forest policy and provided a national legislative and policy framework for the conservation and improvement of forests and the environment. India voluntarily submitted a national forest reference level to the UNFCCC, which included the sustainable management of forests and covered four IPCC carbon pools (above-ground biomass, below-ground biomass, deadwood and litter, and soil organic carbon) and CO_2. The proposed reference level was—49.70 Mt CO_2 eq. and was based on the historical time series (1994–2008). India also

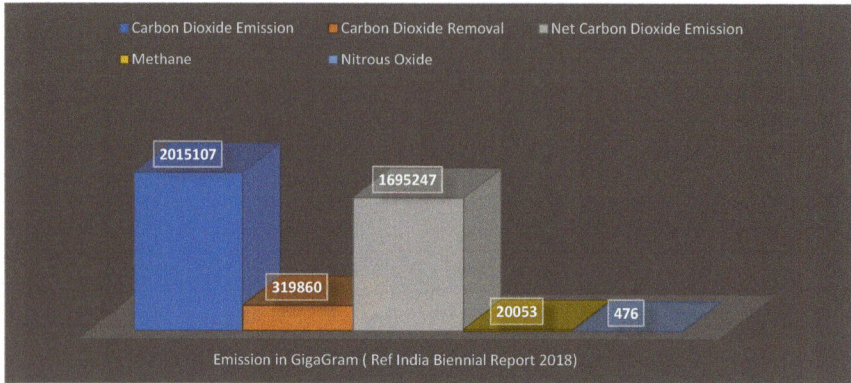

Fig. 4.11   GHG emission of India (GigaGram). (MoEFCC 2018)

reported a number of pilot projects for REDD + that aimed to reduce emissions and sequester carbon and to ensure participation and benefit-sharing in communities. LULUCF sequestered 319,860.23 Gg CO_2 emissions in 2014 (Fig. 4.11), which is about 16% of India's total CO_2 emissions. The total carbon stock in forests for 2017 was estimated at 7,083 Mt. The annual increase of carbon stock was 19.50 Mt carbon (71.5 Mt CO_2 eq.).

For the agriculture sector, mitigation actions included 290 registered CDM projects. Agriculture contributed 16.2% of the total GHG emissions in India and 14.82% to the country's gross value added (at constant 2011–2012 prices). Several initiatives in the sector promoted sustainable development, mitigation, and adaptation. Micro-irrigation resulted in emission reductions of 22.82 Mt CO_2 eq. during the period 2010–2016.

A summary of actions/proposals to mitigate climate action is provided below:

1. India voluntarily pushed up its target of reducing emission intensity of its GDP by 33–35% from 2005 levels by 2030. India's multiple mitigation actions taken between 2005 and 2014 led to emission intensity decline by 21%.
2. India and France jointly initiated the International Solar Alliance (ISA) of 121 sunshine countries to work for efficient exploitation of solar energy to reduce dependence on fossil fuels. Besides, India has partnered with 22 member countries and the European Union in the 'Mission Innovation' on clean energy.
3. The share of non-fossil fuel-based electricity generation installed capacity reached 35.5% in June 2018 and solar energy capacity increased by about nine times from 2.63 gigawatts in March 2014 to 23.02 gigawatts in June 2018. Overall share of renewable energy continued to progressively increase in the electricity mix and in the year 2017–2018, renewable energy generation in India crossed 100 billion kilowatt-hour.
4. Energy efficiency—India launched an ambitious plan[9] to replace all incandescent lamps with light emitting diode (LED) bulbs leading to energy savings of

---

[9]Ujala Scheme.

up to 100 billion kilowatt hours (kWh) annually. More than 312 million LED bulbs were distributed that led to a reduction of 33 $MtCO_2$ emissions per year till October 2018. Around 2 million energy-efficient fans were distributed, which resulted in an estimated energy savings of 191.41 million kilowatt-hour per year and GHG emission reduction of 0.1 $MtCO_2$ per year.

5. A total of 66 super-critical thermal units with a total capacity of 45,550 megawatts were installed that reduced CO_2 emission by about 7 million tons in 2016–2017.

6. A retrofitting project was initiated by Energy Efficiency Services Limited (EESL) in 2014 in the building sector that led to energy savings of 79.8 GWh and CO_2 reduction of 65,578 tons till October 2018.

7. More than 50 million LPG connections were provided to below poverty line (BPL) families.

8. The national policy on biofuels adopted in 2018 aims to increase the usage of biofuels in the energy and transportation sectors of the country during the coming decade. Currently, biodiesel blending percentage in diesel is less than 0.1% and ethanol blending percentage in petrol is around 2%. An indicative target of 20% blending of ethanol in petrol and 5% blending of biodiesel in diesel is proposed by 2030.

9. The national electric mobility mission 2020 (NEMMP) is a significant and ambitious initiative aimed at gradually ensuring 6–7 million electric/hybrid vehicles in India by the year 2020.

10. Almost a third of railway network has been electrified (23,555 km by March 2016), and many mass-transit and urban transport projects have also been initiated under the National Urban Renewal Mission.

11. A crop diversification program has been in the states of Punjab, Haryana, and western Uttar Pradesh since 2013–2014 to divert the area under water-intensive paddy to alternative crops like pulses, oilseeds, maize, cotton, and to agro-forestry plantation with the objective of tackling the problem of declining soil fertility and depleting water table in these states. An emission reduction of 0.21 $MtCO_2e$ (2010–2016) has been achieved under this project.

12. Rice intensification and direct seeded rice systems in various regions of the country led to an emission reduction of 0.18 $MtCO_2$ between 2010 and 2016 and 0.17 $MtCO_2$ between 2014 and 2016, respectively.

13. The total carbon stock in the forests for the year 2017 was estimated to be 7083 Mt.

14. A program for planting trees in 8.39 million hectares has been envisaged cleaning the river Ganges program with a potential to sequester around 87.26 $MtCO_2$ per year. A new policy for highways (Green Highway Policy 2015) was launched to promote the greening of national highway corridors across the country with the participation of the community, farmers, NGOs, private sector, institutions, government agencies and the state forest departments. Under this policy, around 140,000 km of national highways will be planted up with trees.

4.7 Climate Vulnerability

India's rapid population, economic and industrial growth has created enormous pressures on its resources. Out of the total geographical area of 3.28 million square km, about 0.27 million hectares is prone to cyclones and 40 million hectares is prone to floods. The coastal areas are most vulnerable to natural disasters as well. The entire east coast of India, the Gujarat coast along the west coast, and the islands of Lakshadweep and Andaman and Nicobar face frequent cyclonic conditions which sometimes cause large-scale destruction of life and property. The super cyclone of 1999 caused massive destruction along the coast of Orissa and its impact was felt several kilometers inland. The tsunami on December 26, 2004 was one of the most serious and unexpected natural catastrophes that impacted life and property located along the coast of Andaman and Nicobar, Tamil Nadu, Pondicherry, and Kerala. State-wise details of damage due to unnatural events such as cyclonic storms, heavy rains, floods, landslides, and earthquakes during the years 2013–2014 to 2017–2018 are provided in Figs. 4.12, 4.13, 4.14, and 4.15.

Extreme conditions and frequent disasters have greatly eroded developmental gains and caused heavy loss of human lives mainly due to fast and unplanned economic development. As compared to inland areas, coastal region is subjected to ecological stress and environmental conflicts due to erosion, overexploitation of natural resources, loss of biological cover, increased salinity, ground water depletion, and pollution. With continuously increasing population pressure in the coastal states (Fig. 4.16), the vulnerability to frequent and intense natural and manmade hazards is on the rise. Analysis of cyclones on the east and west coasts of India between 1981 and 1990 shows that nearly 262 cyclones occurred in a 50 km wide strip causing massive destruction of life and property.

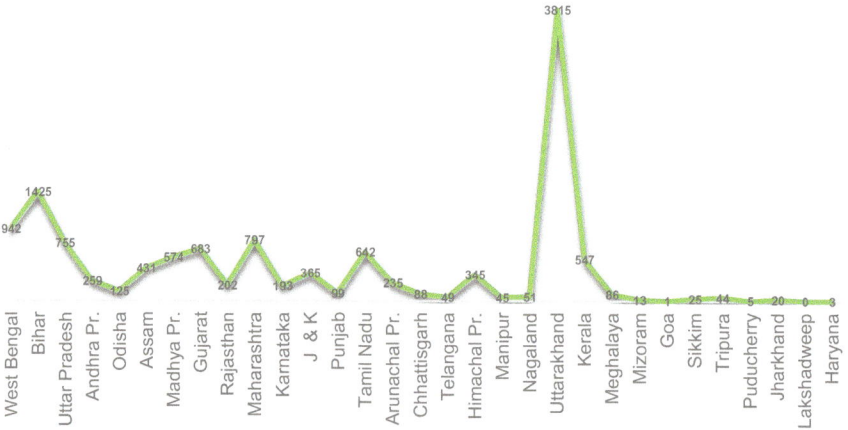

Fig. 4.12 Human lives lost (No.) between 2013–14 and 2017–18. (*Source* Sabha, Lok. "Lok Sabha Question 624" (2018))

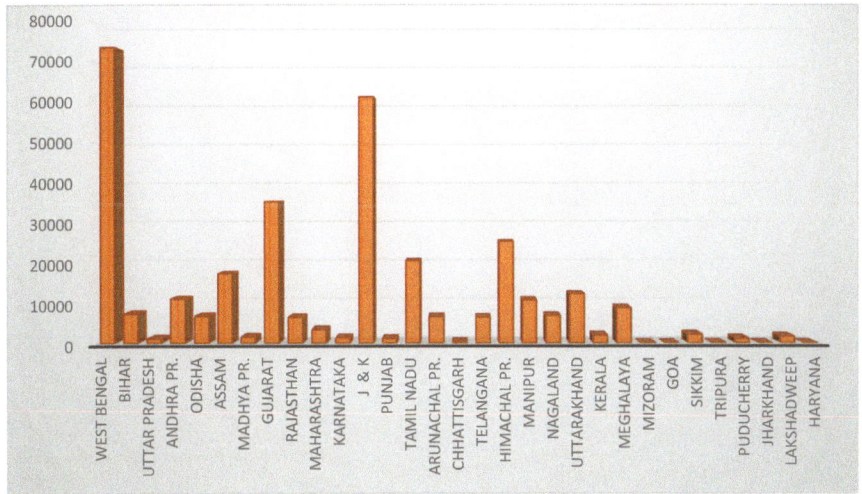

Fig. 4.13 Livestock lost (No.) between 2013–14 and 2017–18. (*Source* Sabha, Lok. "Lok Sabha Question 624" (2018))

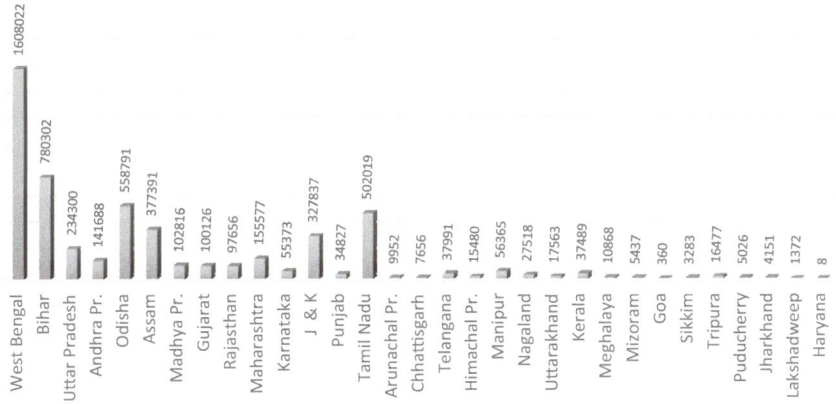

Fig. 4.14 Houses Damaged (No.) between 2013–14 and 2017–18. (*Source* Sabha, Lok. "Lok Sabha Question 624" (2018))

The Indian coast has undergone several changes many of which are related to hazard induced factors including perturbations and stress caused by natural and manmade factors, respectively. The damages caused by these events are mainly due to three factors, namely strong winds, storm surges, and heavy prolonged rainfall. Strong winds cause damage to infrastructure, dwellings, communication systems, and trees, resulting in loss of human life and property. Storm surge causes inundation of low-line areas of coastal region drowning human beings and livestock, eroding beaches and embankments and destroying vegetations, reducing soil fertility, and

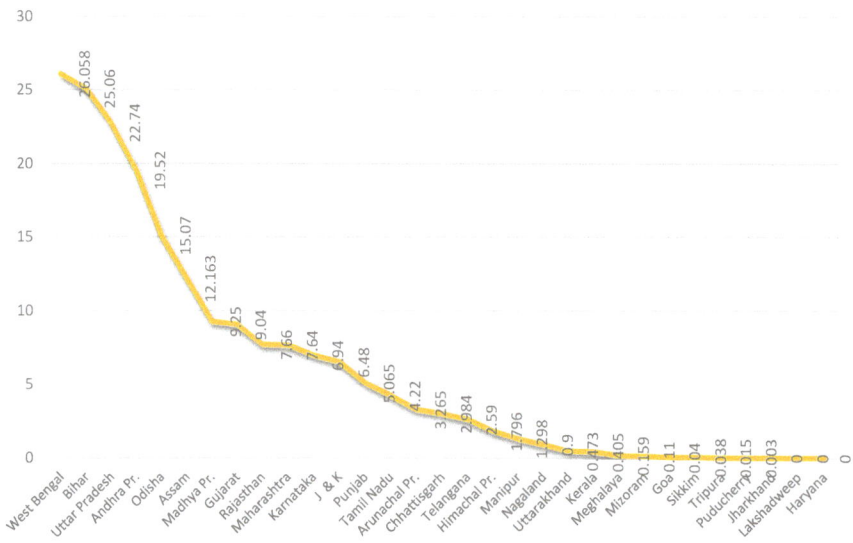

Fig. 4.15 Crops area Damaged (in lakh ha.) between 2013–14 and 2017–18. (*Source* Sabha, Lok. "Lok Sabha Question 624" (2018))

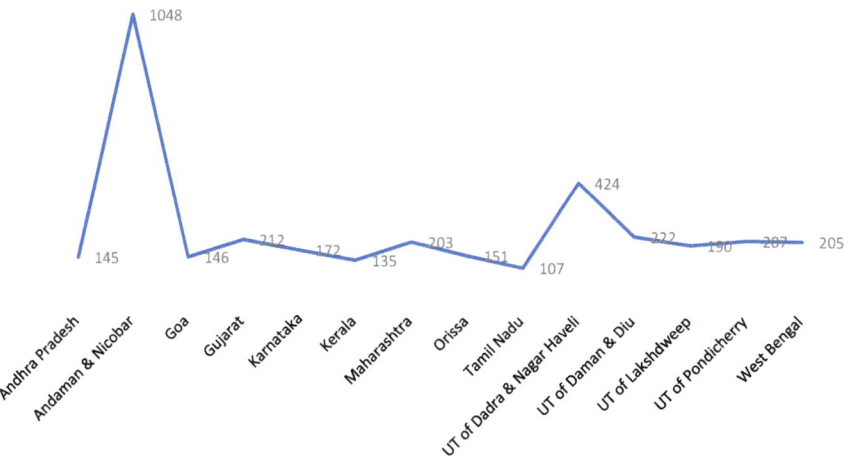

Fig. 4.16 % Increase in coastal population between 1951 and 2001. (*Source*—Chandramouli 2011)

polluting drinking water. Heavy and prolonged rains due to cyclones cause river floods and submergence of low-line areas leading to loss of life and property. Out of 1474, 964 cyclones that originated in the Bay of Bengal and the Arabian Sea during the period of 1877–1990 crossed the Indian coastline inflicting damages to life and property. Expert estimate suggests nearly 5.15 depressions, 1.93 storms and 1.35 severe storms cross the Indian coastal line an average per year.

The current pace of economic growth present enormous challenge as risk factors related to environmental degradation, urbanization, motorization, and industrialization will continue to overwhelm Indian cities and suburbs if steps are not taken to mitigate the growing negative impacts. Many coastal stretches are overpopulated, highly polluted with municipal waste from urbanization and tourism, waste generated from industry, chemicals from fertilizers and pesticides, and silt from degraded catchment. Untreated sewage and other non-industrial waste far exceed the pollution from industrial effluents. Mining of sand results in increased turbidity in the ambient waters which affect benthic organism and primary productivity. Coastal aquaculture mainly related to shrimp farming in some parts has placed considerable pressure on coastal resources. Construction of breakwaters alters the sediment transport mechanism in coastal areas, thereby causing erosion and accretion. The economic liberalization policy of India has encouraged investments over billions of USD per year involving construction of platforms, pipelines, and other structures which are critical component of national economy. However, economic development brings in new threats making the terrestrial and aquatic ecosystems, infrastructure, and population vulnerable to natural and man-made hazards.

The coastal communities in many states are economically weak (*low per capita income, near zero savings, high concentration of population in farming and fishing, high proportion of expenditure on food, low volume of trade per capita, poor credit facilities*); suffer from malnutrition, inadequate sanitation, poor housing and health. Employment is mostly seasonal, and alternative sources of income are limited and unsustainable as the markets for marine and coastal products are disorganized, exploitative, and underdeveloped. The net effect is that coastal communities are highly vulnerable especially during and in the aftermath of a natural disaster.

Rainfall in India is unevenly distributed and provides 70–96% of freshwater replenishment during monsoon season. But the ever-increasing ecological footprint of around 1.3 billion people makes this water insufficient to meet food, industrial, waste recycle, and other requirements. Freshwater has, thus, been exploited since the beginning of twentieth century resulting in shrinking or even drying up of many water bodies (both ground water and surface water bodies). At many places, the underground water is recharged through treated wastewater from sewage and industries that eventually ends up in serious water quality problems. In addition to the above water-related stress, the rainfall observations between 1901 and 2004 indicate a significant positive trend (6% per decade) in the frequency of extreme rainfall events,[10] that is, more than 100 mm per day whereas light and moderate rainfall events show decreasing trend over India (Rajeevan et al. 2008). Frequency and duration of rainstorms have also increased during the past 60 years. This is a clear indication that Indian coast will have to face climate-related hazards in the coming years.

---

[10](Rajeevan et al. 2008).

SECTOR	ENERGY CONSUMPTION (ToE/MT)		
	YEAR 2005	YEAR 2010 (% saving over 2005)	YEAR 2017 (Projected % saving over 2005)
CEMENT	0.08	0.075 (6%)	0.070 (9.63%)
PUPL & PAPER	0.78	0.72 (7%)	0.67 (29.49%)
IRON & STEEL	0.70	0.66 (6%)	0.63 (10.29%)
FERTILIZER	0.63	0.59 (6%)	0.57 (11.11%)

Fig. 4.17 Initial achievement under PAT (CEA. National Electricity Plan 2018)

4.8 Initiatives by Government[11]

National action plan on climate change has set up eight missions to fulfil India's commitment to UNFCCC. One of these eight missions, the national mission on enhanced energy efficiency has spelt the following four initiatives to enhance energy efficiency in energy-intensive industries:

(1) Perform, achieve, and trade scheme (PAT) is a regulatory instrument to reduce specific energy consumption through a tradeable certification process for excess energy saved. The details of initial achievements under PAT are given in Fig. 4.17.

(2) Market transformation of energy efficiency (MTEE) for accelerating switchover to energy-efficient appliances in designated sectors. Under MTEE, two programs, Bachat Lamp Yojana (BLY) and Super-Efficient Equipment (SEEP), have been developed. BLY is a public–private partnership program to accelerate market transformation in energy-efficient lighting. Under this program, over 29 million incandescent bulbs have been replaced by CFLs. SEEP is a program for super-efficient appliances by providing financial stimulus innovatively at critical point/s of intervention. Under this program, ceiling fans were identified as the first appliance to be adopted. SEEP for ceiling fans aims to leapfrog to an efficiency level which will be about 50% more efficient than market average by providing a time-bound incentive to fan manufacturers to manufacture super-efficient (SE) fans and sell the same at a discounted price. The goal is to support the introduction and deployment of super-efficient 35 watts ceiling fans, as against the current average ceiling fan sold in Indian market with about 70 watts rating. For the 12th plan, it is targeted for deployment of 2 million fans during the plan period with an outlay of INR 100 crores. Under this program, maximum of INR 500 per fan incentive will be given to fan manufacturers for manufacture and sale of fans meeting SEEP specifications (CEA, National Electricity Plan 2018).

(3) Energy efficiency financing platform (EEFP) for creation of mechanisms to assist in financing demand-side programs in all sectors by capturing future energy savings.

(4) Framework for energy-efficient economic development (FEEED) for development of fiscal instruments to promote energy efficiency.

---

[11](*Source* CEA, National Electricity Plan 2018).

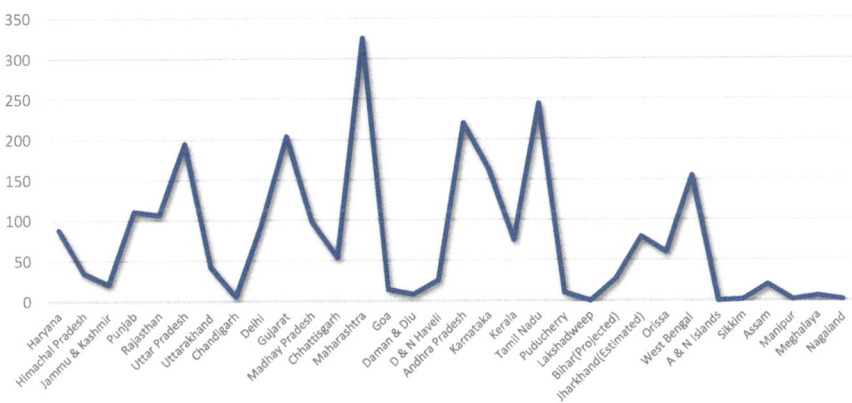

Fig. 4.18 State-wise energy savings targets (cumulative) between 2017–18 and 2026–27. (*Source*— CEA, National Electricity Plan 2018)

Government of India has taken other measures as well, including adoption of super-critical technology for thermal power with 2–3% higher efficiency than the present 500 megawatts sub-critical plants. The national average thermal efficiency of coal/lignite power plants has increased from 32.53% in 2009–2010 to about 34% in 2013–2014. It is expected that the efficiency of coal-based generation would further improve in the period 2017–2022 due to commissioning of large size super-critical units. Old and inefficient units of thermal power stations will be replaced with more efficient units. A capacity of 2398 and 5082 megawatts have been retired during 11th and 12th plan, respectively. All state governments have been requested to save energy between 2017–2018 and 2026–2027 (Fig. 4.18) with targets for each year allocated to monitor their annual performance.

References

Assembly UNG (1982) World charter for nature. United Nations General Assembly Resolution 37(7)

Brundtland GH (1987) Our common future—Call for action. Environ Conserv 14(4):291–294

Central Electricity Authority (2018) National electricity plan (No. Volume 1: Generation). Ministry of Power, Government of India. (https://www.cea.nic.in/reports/committee/nep/nep_jan_2018.pdf)

Chandramouli C, General, R (2011) Census of India 2011. Provisional population totals. Government of India, New Delhi, pp 409–413

Chaturvedi RK, Tiwari R, Ravindranath NH (2008) Climate change and forests in India. Int For Rev 10(2):256–268

Fleming JR (1999) Joseph Fourier, the 'greenhouse effect', and the quest for a universal theory of terrestrial temperatures. Endeavour 23(2):72–75

Fleming J (2013) The callendar effect: the life and work of guy stewart callendar (1898–1964). Springer Science & Business Media

Ghosh S (2020) Moving away from state and capital. Resistance and Grassroots Solutions, Climate Justice and Community Renewal

Government of India (2008) National action plan on climate change. Prime Minister's Council on Climate Change

Gray V (2013) The greenhouse revisited

India MoE (2004) India's national communication of the UNFCCC. Ministry of Environment and Forests, Government of India, Delhi

Joos F (1996) The atmospheric carbon dioxide perturbation. Europhys News 27(6):213–218

Keeling CD (1960) The concentration and isotopic abundances of carbon dioxide in the atmosphere. Tellus 12(2):200–203

Krishnamurthy V (2011) Extreme events and trends in the Indian summer monsoon. Center of Ocean-Land-Atmosphere Studies

Latha SS, Prasad MBK (2010) Current status of coastal zone management practices in India. In: Management and sustainable development of coastal zone environments. Springer, Dordrecht, pp 42–57

MoEFCC G (2015) India: first biennial update report to the united nations framework convention on climate change

MoEFCC (2018) India: second biennial update report to the united nations framework convention on climate change. Ministry of Environment, Forest and Climate Change, Government of India

Pandve HT (2009) India's national action plan on climate change. Indian J Occup Environ Med 13(1):17

Parikh JK, Parikh K (2002) Climate change: India's perceptions, positions, policies and possibilities. Indira Gandhi Institute of Development Research, Mumbai

Pradhan S, Goswami G (2019)India's second biennial update report: five key takeaways. Curr Sci (00113891)

Rajeevan M, Bhate J, Jaswal AK (2008) Analysis of variability and trends of extreme rainfall events over India using 104 years of gridded daily rainfall data. Geophys Res Lett 35(18)

Sabha L (2018) Lok sabha question 624

Sikka P (2012) Climate change, India in focus: mitigating impacts of global warming. Uppal Publishing House

Srivastav A, Srivastav, Nishida (2019)The science and impact of climate change. Springer, Berlin

Stern DI, Jotzo F (2010) How ambitious are China and India's emissions intensity targets? Energy Policy 38(11):6776–6783

Subramanian A, Birdsall N, Mattoo A (2009)India and climate change: some international dimensions. Econ Polit Wkly 43–50

Summit E (1992). Agenda 21. The united nations programme for action from Rio

UNFCCC website https://unfccc.int

Zhu B, Bin Su, Li Y (2018) Input-output and structural decomposition analysis of India's carbon emissions and intensity, 2007/08–2013/14. Appl Energy 230:1545–1556

Chapter 5
Fourth Industrial Revolution and India

Abstract The world has now moved into the fourth phase of industrial revolution—the digital revolution where either knowledge workers, or robots, or a combination of both will produce most goods and services by focusing on creativity. The revolution offers a great opportunity to all countries to usher into a more sustainable energy management system that is environment-friendly such as solar and wind. It will be counterproductive for India to remain thirsty for oil and coal irrespective of the perceived devastations both environment and economic. With the demand for electricity alone expected to cross 950 gigawatts by 2030, India will have to commit huge upfront investment in energy sector and human skills to remain competitive economy.

Keywords Fourth industrial revolution · Carbon emission · Lithium cell · Global warming · Carbon intensity · Fuel cell · Fusion energy · Fuel wood · Biomass fuel · Fuel value · LED

5.1 The Fourth Industrial Revolution

The term 'Fourth Industrial Revolution' was coined by Dr. Klaus Schwab, founder, and Executive Chairman of World Economic Forum. In his article titled 'The fourth industrial revolution—what it means and how to respond' published in Foreign Affairs,[1] December 12, 2015, he mentioned that:

> We stand on the brink of a technological revolution that will fundamentally alter the way we live, work, and relate to one another. In its scale, scope, and complexity, the transformation will be unlike anything humankind has experienced before … There are three reasons why today's transformations represent not merely a prolongation of the third industrial revolution but rather the arrival of a fourth and distinct one: velocity, scope, and systems impact. The speed of current breakthroughs has no historical precedent. When compared with previous

---

[1] *Source* https://www.foreignaffairs.com/articles/2015-12-12/fourth-industrial-revolution and Schwab, Klaus Martin. 'The fours industrial revolution [E-source].' *KlausMartin Schwab Available at:* https://www.foreignaffairs.com/anthologies/2016-01-01/fourth-industrial-revolution.

© Springer Nature Singapore Pte Ltd. 2021   147
A. Srivastav, *Energy Dynamics and Climate Mitigation*,
Advances in Geographical and Environmental Sciences,
https://doi.org/10.1007/978-981-15-8940-9_5

industrial revolutions, the fourth is evolving at an exponential rather than a linear pace. Moreover, it is disrupting almost every industry in every country. And the breadth and depth of these changes herald the transformation of entire systems of production, management, and governance.

Most of us are conversant with the first industrial revolution that used coal and water and generated steam power to mechanize production. This was followed by second phase during late nineteenth century to generate electric power for mass production. The third phase which started in early twentieth century used electronics and information technology to automate manufacturing. The world has now moved into the fourth phase of industrial revolution—the digital revolution that started sometimes during the middle of the last century. The fourth revolution is characterized by digital tools that generate, store, analyze, communicate, and assist in decision making as and when required. Computers, super computers, mobile phones, drones, virtual assistants, driverless cars, and robots have blurred the lines between the physical and non-physical world (Box 1). Digital fabrication technologies interact with scientists, doctors, engineers, and others for computational design, manufacturing, materials engineering, and synthetic biology.

Box 1 —Different phases of industrial revolution
[Adopted from https://www.foreignaffairs.com/articles/2015-12-12/fourth-industrial-revolution]

Period	Transition Period	Energy Resource	Main Technical Achievement	Means of Transport
1760–1900	1860–1900	Coal	Steam Engine	Train
1870–1940	1940–1960	Oil Electricity	Internal Combustion Engine	Train, Car
1930–2000	1980–2000	Nuclear Energy, Natural Gas	Computers, Robots	Car, Plane
2000 and beyond	2000–2010	Green Energies	Internet, 3D Printer, Genetic Engineering	Electric and driverless cars, Ultra-Fast Train, Drones, Air Taxis

The first two industrial revolutions advocated meeting the energy supply, disregarding environmental harmony, sustainability, equity, and social justice. It was based on the premise that economic growth was directly linked to energy consumption (Box 2) and therefore, ignored energy conservation, energy efficiency, environment, and health impacts.

Box -2

Development ∝ Economic Growth ∝ Energy Consumption ∝ Energy Demand Projection ∝ Energy Supply Increase

Photo-Author (London Science Museum)

The Second
Industrial
Revolution,
1870–1914

The second half of the nineteenth century brought a
huge range of new techniques and new materials to
the industrialised nations. To Alfred Krupp it was the
'Age of Steel.' However, it was also the age of chloroform,
electrical power and high explosive. Commentators
came to speak of a 'second industrial revolution.'

The first industrial revolution had been characterised
by the use of coal, steam and iron. Technology now
became more complex and more diverse. The most
significant developments in this new age were in the
chemical and electrical industries. Synthetic dyes,
artificial fertilisers, plastics, new textiles, and drugs
such as aspirin, were invented. Life at home and in
the factory were transformed by the development
of electricity for power, heat and light. Electricity and
chemistry also had an enormous cultural impact by
making possible the new media of cinema and radio.

Many of these innovations came from Germany,
France and America and by 1870, Britain 'the first
industrial nation,' was overtaken in industrial output
by the USA. This process started the discussion of
British 'decline' which continues today. Nevertheless,
British production, national income and living
standards continued to rise.

The conventional paradigm changed after many catastrophic global, regional, and local events including the First and Second World Wars, the oil crisis of 1970s, innumerable nuclear accidents in nuclear power plants, destruction of forests, environmental degradation due to large hydro projects, and the evidence of global warming. The developed countries were compelled to change their approach to energy generation which was based on the fact that the performance of energy using equipment was more significant than the quantum of energy used. For example, improving the cooling efficiency of air conditioners to reduce electricity consumption or improving the heating efficiency of electric heater or the efficiency in mobility provided by vehicles that use gasoline to convert heat energy to mechanical, and so on. In other words, it was the energy efficiency that took precedence over all other considerations. The new paradigm, while agreeing with the prevailing approach of substantial

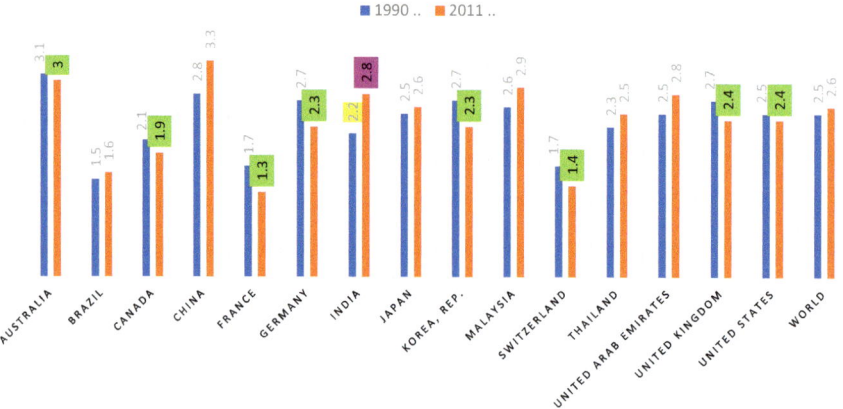

Fig. 5.1 Changes in carbon dioxide intensity-carbon dioxide emission (KGS) per kilogram of oil equivalent energy use (World Bank 2015. *World Development Indicators 2015*)

enhancement of energy supply and consumption for economic development, clearly recognized that inefficiency in energy management was the largest source of carbon emission that was responsible for increasing global temperature with consequent impacts. Accordingly, the new paradigm gave precedence to energy efficiency and CO_2 intensity (Fig. 5.1) over magnitude of energy consumption and encouraged all countries to innovate and introduce energy efficiency measures in energy generation, transmission, and end-use.

The oil crisis of 1970s was a clarion call for the developed nations and many of them switched over to improved ways of energy generation and distribution quickly replacing archaic technology. But the developing and underdeveloped world could not keep pace with this transition as most of them were heavily dependent on industrialized nations for energy sources and technology. Instead, the industrialized countries happily passed on their obsolete technology disguised as international aid to the poor nations. The energy generation and consumption pattern in different countries eventually led us to a 'energy-temperature trap', whereby rising temperatures increased the demand for cooling, and increased cooling led to rising temperatures. A case in point is the proliferation of coal-based thermal power plants and rampant increase in vehicles using petrol and diesel. This eventually results in rapid increase in GHG emission and consequent rise in atmospheric temperature that compels consumers to use air conditioners in homes, offices, and vehicles. While this cools the interior of buildings and vehicles, it generates massive heat in the peripheral atmosphere, most of which cannot escape upwards[2] and thus people are forced to use more air conditioners. The net effect is—tremendous increase in energy consumption most of which is produced by burning coal and oil—a perfect case of double-whammy.

---

[2]High-rise building, road surface re-radiate infrared radiations repeatedly heating up buildings and roads. Absence of avenue trees add to the woes.

Under the current scenario when the threat of annihilation by climate-related disasters in a reality, optimizing energy demand, reducing carbon emissions by switching over to renewable sources of energy, and improving efficiency[3] of energy both on the supply and demand side will have tremendous developmental benefits. Shift to renewable may take time due to high cost involved in conversion to electricity. In the meantime, countries can improve supply-side efficiency in burning of oil, gas, and coal as well as production, transmission, and distribution of electricity. For example, improving efficiency of sub-critical coal-based thermal power plants can prevent 35% losses due to wastage in transmission, distribution, and voltage adjustment.[4] Similarly, improving the engines and designs of vehicles, design, and type of construction material for buildings, and manufacturing sector changes can help in demand-side energy optimization. Most of the demand for illumination, heating, and cooling for human comforts, be it in cars, buses, houses, offices, and factories, can be optimized through better design, space management, judicious placement of doors and windows, passive solar heating, insulation, ventilation, equipment, and time management. Large complexes such as shopping malls, railway stations, and airports can make use of sunlight and air circulation by simple modifications in design. Such buildings can preferably adopt 'zero-energy, zero carbon' code, meaning thereby that they will produce their own energy on-site, preferably through renewable sources and emit no CO_2. Simple actions such as trees planting around such buildings and along highways and motorways can bring down temperature by a few degrees that can substantially cut down cooling requirement of such complexes.

In addition to improving energy efficiency,[5] it is important to reduce the energy-related carbon dioxide emission that is directly dependent on total energy consumption and carbon intensity.[6] The carbon intensity varies with the source of energy generation such as coal, gas, and petroleum. An energy policy that reduces energy intensity[7] by increasing energy efficiency through low-carbon lifestyle can substantially reduce carbon emissions. This is easier stated than achievable because global economy is poised to quadruple by 2050 and the current emission trends suggest a potentially catastrophic trajectory for carbon dioxide leading to 5 °C increase in temperature (compared to preindustrial period). Developing nations discredit the developed countries for consuming five times more energy per capita (releasing almost two-thirds of energy-related carbon dioxide) and the developed countries blame the developing world for archaic technology.

---

[3] Energy efficiency reduces energy bills for consumers, increases the competitiveness of industries, and creates jobs.

[4] China has increased the energy efficiency of coal-fired thermal plants by 15% by introducing supercritical and ultra-super-critical technologies. Energy efficiency measures adopted by the USA in 1970s saved around $365 billion in 30 years' time.

[5] Energy efficiency is essential for the 2 °C trajectory. In the short term, the largest and cheapest source of emission reductions is increased energy efficiency on both the supply and demand side in power, industry, buildings, and transport.

[6] Carbon intensity is defined as the units of CO_2 produced by a unit of energy consumed.

[7] Energy intensity is defined as energy consumed per dollar of gross domestic product.

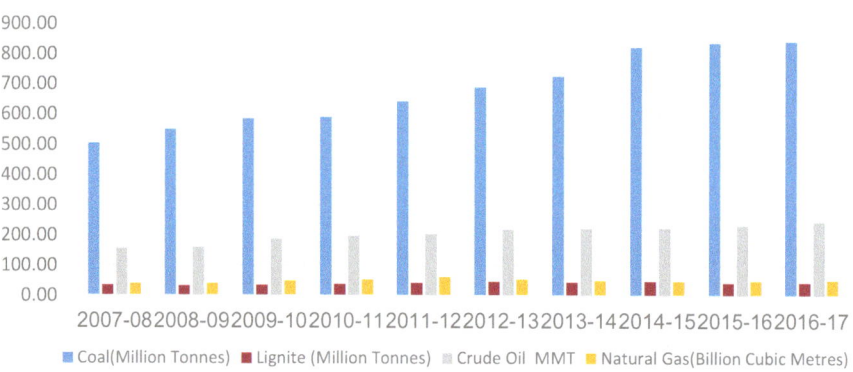

Fig. 5.2 Trends in consumption of energy sources in India (*Source* Energy Statistics, Statistics Energy 2018)

It appears that we have already crossed the threshold levels and even if the carbon dioxide emissions between 2000 and 2050 are restricted to 1000 billion tons, there is a possibility of global temperature exceeding 2 °C. Ironically, among the four principal polluters (i.e. India, China, USA, and Russia), USA has committed to reduce carbon emission by 26–28% from 2005 levels by 2030 (anticipated actual emission would be 4500 million tons in 2030). Thermal power plants in the USA contribute nearly a third of GHGs, and therefore, it will have to switch from coal to gas/renewable by 2030 (Allen et al. 2009). On the other hand, China's CO_2 emission is expected to reach its peak of 18,000–20,000 million tons by 2030 from the ongoing 9000 million tons[8] and India is expected to reach 4000–5000 million tons from its ongoing CO_2 emission of 2000 million tons by 2030. When will CO_2 emissions of India and China decline is a serious question being asked by global scientific community, especially considering the fact that India is embroiled in employment generation, industrialization, and economic deceleration? Under the circumstances, investment in energy-efficient technologies and innovations will take the back seat. This in turn means delay in phasing out use of coal (Fig. 5.2) and oil-based energy production and investing in low-carbon/zero-carbon system for industry, buildings, and transport. At the same time, addressing the galloping and wasteful use of energy by rich and upper middle class is by no means a humongous effort that requires compromises in aspirations, consumption pattern, and lifestyles that many of us are unwilling. For example, how many are willing for 'no use' of AC while traveling in cars or using public transport or using bicycles for commuting or using solar cookers for making food?

If the global warming must stay below 2 °C then both developed and developing countries will have to increase efficiency and decrease carbon intensity at the same time. Arresting this process of CO_2 and temperature rise, the global community must:

---

[8] In 2012, China was producing 758 gigawatt (out of total production of 1145 gigawatt) for coal-based thermal power plants.

a. Shift the energy mix from fossil fuel to renewable sources.
b. Make renewable energy cost less than non-renewable energy to make it globally acceptable.
c. As an alternative, increase the cost of non-renewable energy.
d. Motivate/compel urban population to adopt energy mix with greater share of renewable, especially solar water heaters, solar cookers, and solar lamps.
e. Encourage/compel commuters to use public transport system.
f. Subsidize low/zero carbon emission fuels.

At the same time, there is a need to shift emphasis from the irresponsible consumption of energy to improved provision of energy services, and the true indicator of the sustainable energy development should be the level of energy services enjoyed by the population, particularly the rural poor, along with the magnitude of per capita energy consumption. This is undoubtedly a difficult task for many countries, including India that will continue to use conventional sources including biomass-based fuels irrespective of the climate consequences (Srivastav 2019). Poverty will compel people not to pay high cost for renewable energy, and the current economic collapse (COVID-19 impact) will only add fuel to the fire with higher living cost, massive jobs cut, reduced wages, and escalating inflation. Energy reforms, even in the developed world will take the back seat and, in all likelihood, Germany will continue to use coal despite current EU policy of 20-20-20 (Musall and Onno 2011). Similarly, France will continue to be committed to nuclear energy, and Britain goes all out for shale gas and nuclear energy. All indicators point to the fact that use of conventional energy will continue unabated till the cost of energy from renewable sources such as wind and solar remains higher than nuclear, coal, and gas generated electricity.

Fourth industrial revolution is evolving differently from its predecessors in the sense that either knowledge workers or robots or a combination of both will produce most goods and services by focusing on creativity and achieving organizational goals more efficiently. In simple terms this will mean super-skilled humans will produce goods and services with minimum energy. The revolution thus offers a great opportunity to all countries to usher into a more sustainable energy management system that is environment-friendly. With the introduction of solar cells and high storage lithium batteries, the transportation system will see incredible revolution, coal and oil mining will be history, manufacturing will witness 90–100% automation and household/building energy management will be highly energy-efficient. With these energy forecast, it will be counterproductive for India to remain thirsty for oil and coal irrespective of the perceived devastations both environment and economic. With the demand for electricity alone expected to cross 950 gigawatts by 2030, India will have to commit huge upfront investment in energy sector and human skills to remain competitive economy.

5.2 Readiness for Low/Zero Carbon Growth

It has been more than 30 years that many of us were made conversant with the term global warming. Subsequent scientific and other reports persuaded us to make a sincere and time-bound effort to reduce greenhouse gas emissions. Unfortunately, little worthwhile has been achieved despite international conventions and declarations, laws, policies, and rhetoric. The following factual details support the foregoing argument:

1. The world oil production has been on the rise (Fig. 5.3):
2. The global atmospheric carbon dioxide concentration is on the rise (Fig. 5.4):
3. Carbon dioxide emission from use of coal, oil and gas is on the rise (Fig. 5.5):
4. Global trend in carbon dioxide concentration and sea water pH indicate worsening scenario (Fig. 5.6):
5. Global average temperature is rising (Fig. 5.7):
6. By mid-2090s, the global sea level is projected to reach 0.22–0.44 mm above 1990 levels (Fig. 5.8).

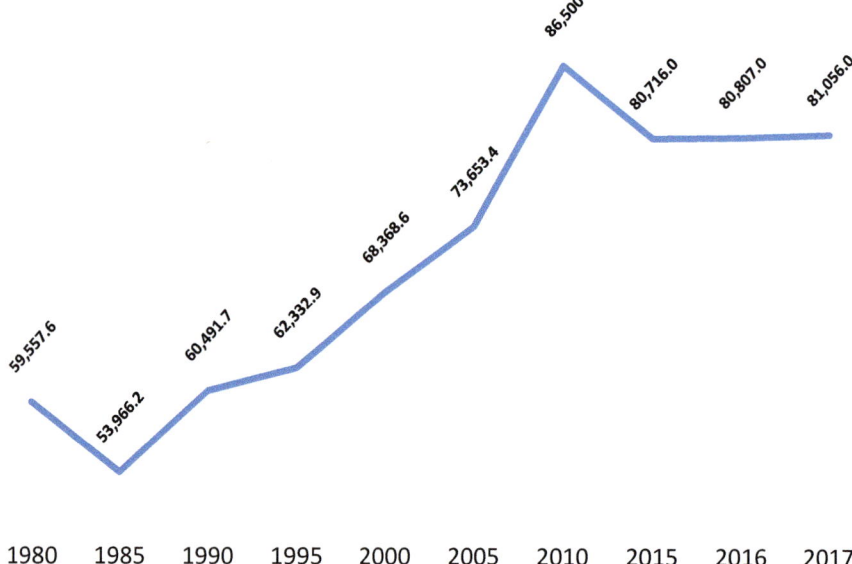

Fig. 5.3 Global crude oil production (Thousand Barrels per Day) (*Source* International Energy Agency. *Key World Energy Statistics* 2018)

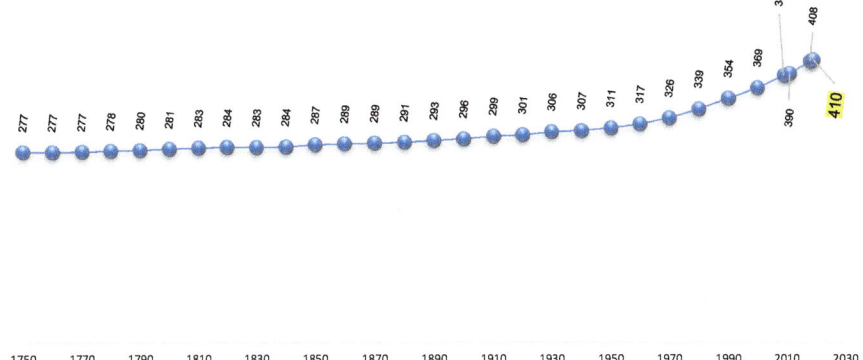

Fig. 5.4 Decadal Changes in carbon dioxide concentration (parts per million by volume) between 1750 and 2019 (*Source* Compiled from Brown (2013) and Website https://www.climate.gov/news-features/understandingclimate/climate-change-atmospheric-carbon-dio)

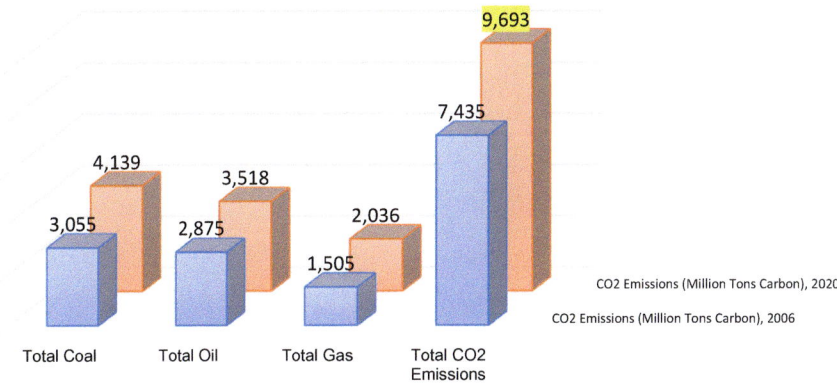

Fig. 5.5 CO$_2$ emission (*Source* Compiled from Brown 2003, 2013; and IPCC 2007)

5.3   Carbon-Based Economies Will Eventually Prove Cataclysmic

Why is it that despite having irrefutable scientific evidences of climate change and its consequences, we have achieved very little so far. One argument is that it is not in the interest of any one country to reduce emissions if others do not follow. Presently, the world is divided between two broad groups of nations: The industrialized and rich developed countries that have generated insurmountable quantities of GHGs in the past and currently have the technology and money to switchover to renewable sources of energy. Past few decades have witnessed some sensitivity by these advanced countries toward climate change mitigation, and appreciable actions have been taken in this direction including large-scale generation of renewable energy, introduction of

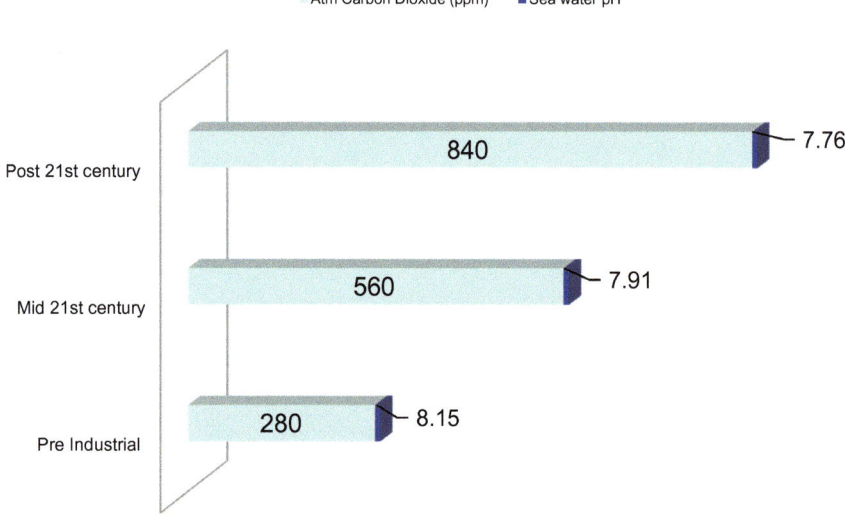

Fig. 5.6 Atmospheric CO$_2$ and Sea water pH (*Source* Seibel and Fabry 2003)

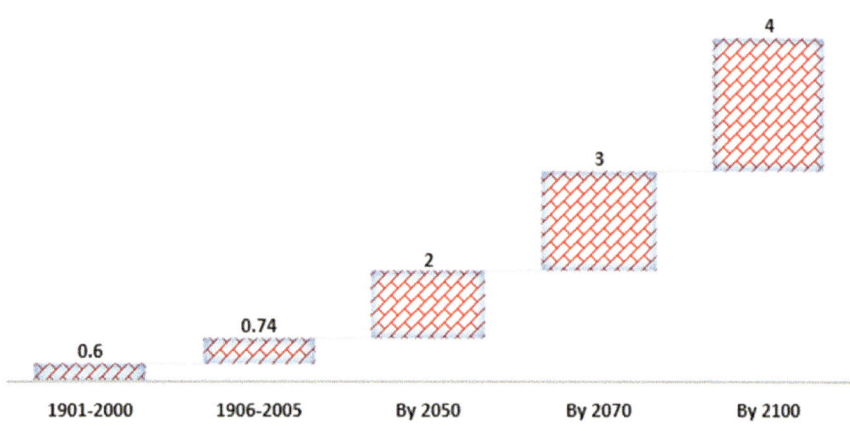

Fig. 5.7 Global average temperature (°C) changes so far and anticipated (*Source* UNFCCC and https://www.un.org/en/sections/issues-depth/climate-change)

electric motor vehicles, switchover to gas-based heating, aggressive use of bicycles, protection of forests, and so on. The second group consists of poor and developing nations who have been dependent on the rich nations for resources and technology. These nations continue to rely on carbon-based economy and constantly seek funds and technology from the rich countries. Fortunately, a new political force emerged in Europe and North America during 1980s that was inclined toward environmental justice and equity and they vigorously pursued environment protection agenda across the globe and eventually succeeded in compelling rich countries to change their

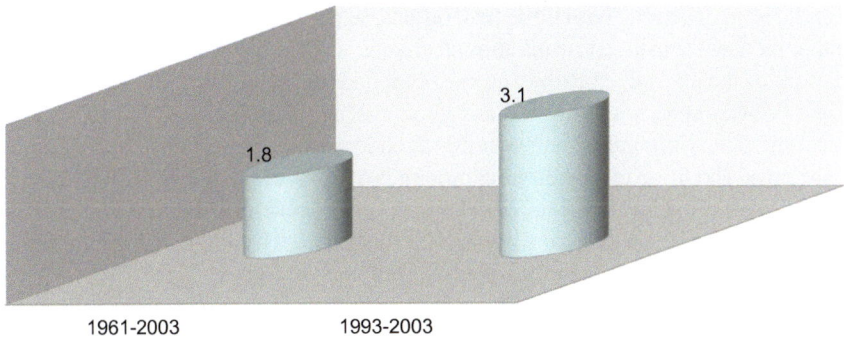

Fig. 5.8 Average sea level rise (in mm) (*Source* IPCC 2007)

energy strategies. Unfortunately, environmentalism had never been the vote bank for the political establishments of developing nations where food and water security, education, health, infrastructure, and employment (and not energy) play decisive role in win/loss of governments.

Politics aside, the science of climate change is clearly of the view that global warming in excess of 2 °C will be extremely challenging and therefore all nations must work toward reducing the carbon intensity of production. Efforts of the pressure groups did result in reducing the carbon intensity from 480 to 380 g CO_2 per USD between 1990 and 2010 (Rozenberg et al. 2015). However, in order to maintain less than 2 °C trajectories, most of the scientific models indicate a carbon intensity between 33 and 73 g CO_2 per USD against the ongoing average of 360 g CO_2 per USD (Box 2). This means radical shift in the ongoing production mechanisms to carbon neutral system.

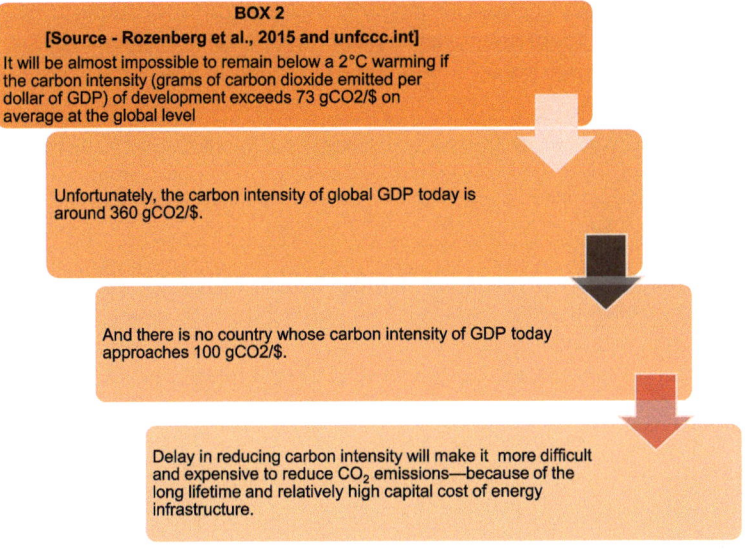

India is no different. With more than 60% rural population and modest education, people are least sensitive toward climate change mitigation. Almost all of them are unaware of the causative factors and the point and non-point sources of greenhouse gases. For example, not many Indians know that the thermal power plants in their cities emit obnoxious gases such as sulfur dioxide, carbon mono and dioxides, oxides of nitrogen in addition to dangerous compounds of arsenic and fly ash. If people living in the vicinity are informed of the health impacts of all residues, there will be massive public support for solar and wind power. But this may not be in the interest of mining and other groups that may not be interested in financial losses.

The economics of thermal power projects includes the cost of mining, transportation, power generation, and distribution, but intentionally excludes the cost related to environmental and climate change impact. This is deliberately done to keep high profit margins for mining and power industries and ensuring viability of project. A win–win situation for all—the producers as well as end-users of electricity—a path of least resistance. Adding the cost of climate change impact will, in most cases, make the project unviable, a decision most economist detest to arrive at. Conservative estimates by International Energy Agency (Outlook, India Energy 2015) indicate that 30% of India's population in 2040 (close to 480 million) mostly in rural areas will remain bereft of clean energy despite the best possible efforts from all. Not only that, more than 60% of India's power will continue to be generated by coal-based thermal plants. Under these circumstances, it will be imprudent on anyone's part to accept that India will or can switch over to low-carbon economy. And if India continues to release greenhouse gases in quantities greater than natural sinks can absorb, the climate will continue changing with greater risks for wellbeing, health, economic development, and poverty reduction. The more we delay actions for low-carbon pathway, the more difficult and expensive it will be to reduce GHG emissions sufficiently to stay within a low budget of cumulative emissions—because of the long lifetime and relatively high capital cost of energy infrastructure.

Most Indians have forgotten what the then UN Secretary General said in 1970[9] while addressing 7th session of the General Assembly in New York:

> As we watch the sun go down, evening after evening, through the smog across the poisoned waters of our native Earth, we must ask ourselves seriously whether we really wish some future universal historian on another planet to say about us: 'With all their genius and with all their skill, they ran out of foresight and air and food and water and ideas.

The young generation has a propensity to use more energy as they earnings grow and to substitute manual work with machines not knowing where the energy comes from, how is it generated, how many million tons of poor quality coal is extracted and burnt, what is the quantum of GHGs released, and what happens to the residues left in the thermal power plants. Unfortunately, the older generations that were brought up in early and mid-twentieth century when energy was scarce and expensive are gradually disappearing. They have, for bad reasons, failed to inculcate the habits of turning off appliances or of walking/cycling rather than driving. Among young generation,

---

[9]Cross-Ref: Nicholson and Sikina 2016.

there is an inclination and money to visit beautiful tourist places in India and abroad, but only a few have time to visit thermal, hydel, and nuclear power stations nor are such visits supported by the governments for fear of criticism. There is a noticeable decline in the environmental and climate movements that we all witnessed in 1960s, 1970s, and 1980s. This attitudinal change is a clear indicator that:

i. Prosperity is directly proportional to desire for replacement of manual work with machines most of them driven by energy derived from fossil fuel.

ii. As the prosperity expands its network, bigger and ostentatious machines will be replaced by smaller more sophisticated ones that are energy efficient.

iii. Prosperity has brought in lavish lifestyle with more cars and two wheelers and less travel by public transport. Poor and inefficient public transport system discourages its use by citizens and therefore individual car/vehicle ownership, though undesirable, is preferred.

iv. The middle class prefers vehicles that are cheap, use diesel/gas, and is low on maintenance.

v. There is a tendency to use old vehicles (more than 15 *years) that devour enormous quantities of fuels.*

vi. Commercial interests invariably have short payback period (5–7 *years) on their investment and no carbon-reducing interventions can be achieved in this period. Moreover, none has time and patience to wait for 50–70 years or for the welfare of posterity.*

vii. Prosperity, propensity, and lack of self-restraint has generated so much waste products that is almost impossible to recycle; that is being burnt regularly to generate obnoxious gases such as dioxins and furans.

viii. Climate mitigation is not about tree planting alone. There is a general belief in India that merely planting trees will minimize climate-related disasters and therefore natural forests can be conveniently diverted for economic development. Governments after governments have announced planting of billions of trees in every state, especially during monsoon, but most of these trees fail to survive.

ix. There is a general belief in India that tree planting is a low-priced activity that can be performed by all and sundry. The truth is tree planting is a highly scientific activity starting with seed collection, storage, transport, germination rate, growth rate, branch angle, canopy size, height and shape, disease resistance, root-shoot ratio, health of soil including moisture, water retention, infiltration rate, carbon, nitrogen and microorganism, overall lifespan, and so on. In some countries the health of each tree is regularly monitored through sophisticated equipment costing millions of dollars, and necessary interventions are made if necessary. In other words, low-priced and unscientific tree planting whether inside natural forest or outside will not sustain the test of time.

x. Except for a few isolated incidents, nowhere in the country there has been a demand by the electorate against stopping global warming actions and activities. On the contrary, there have been many instances where the constituency demanded for diversion of forested lands for personal and societal gains.

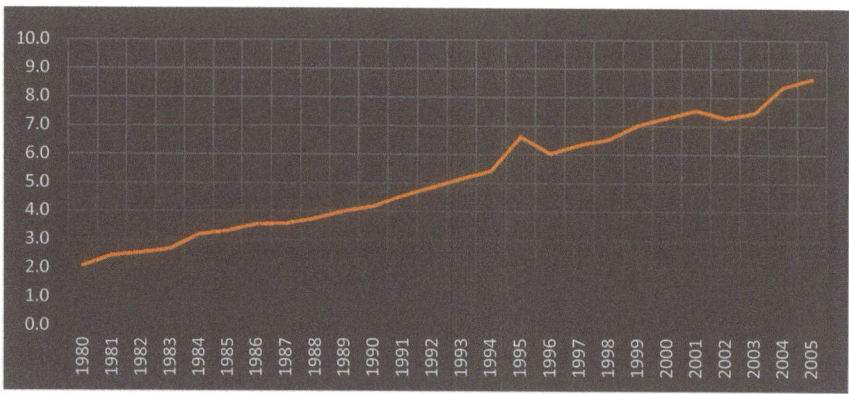

Fig. 5.9   Coal consumption trajectory of India (Quadrillion Btu) (*Source* Brown 2012)

For next 50 years, Indian governance will be under extreme stress for two reasons. One, switching over from non-renewable to renewable sources of energy in such a manner that provides respectable per capita energy to 1.5 billion people to say the least, and second, provide enough energy to manufacturing and export sector to remain stable in global competitive market. This is going to be nerve-wracking since Indian climate mitigation strategies have been conservative with focus on tree plantation and renewable energy primarily on solar and wind, improving energy efficiency, and building climate-resistant smart cities with ambitious targets. With insufficient climate adaptation and mitigation funds, rising unemployment, slowing down of economy, and outmoded technology, one can safely predict that GHG emission will go up in absolute terms. Unlike many advanced countries, India has not announced timeline for closure of coal-based obsolete thermal power plants, reinforcing a carbon-intensive path (Fig. 5.9). The targets set for emission intensity reductions (the amount of greenhouse gases emitted per unit of GDP) imply that absolute emissions would continue to increase and there is no commitment as to when the emissions would peak and eventually reduce.

The World Economic Forum report 'Readiness for the future of production 2018', mentions that the future of production is being transformed through innovation, artificial intelligence, digitization, internet of things, and renewable energy (Kearney 2018). The future of development will depend on speed and scope of innovation and related factors, including readiness of a country to change. The report assesses 100 countries based on two major components – the structure of production and the drivers of production. The framework used for the assessment is provided in Fig. 5.10.

On the basis of their assessment, the countries have been divided into four groups, viz.:

- High potential—Limited current base and positioned well for future. This group included nations that have limited production base currently but have strong drivers of production, indicating that they have capacity to increase production in

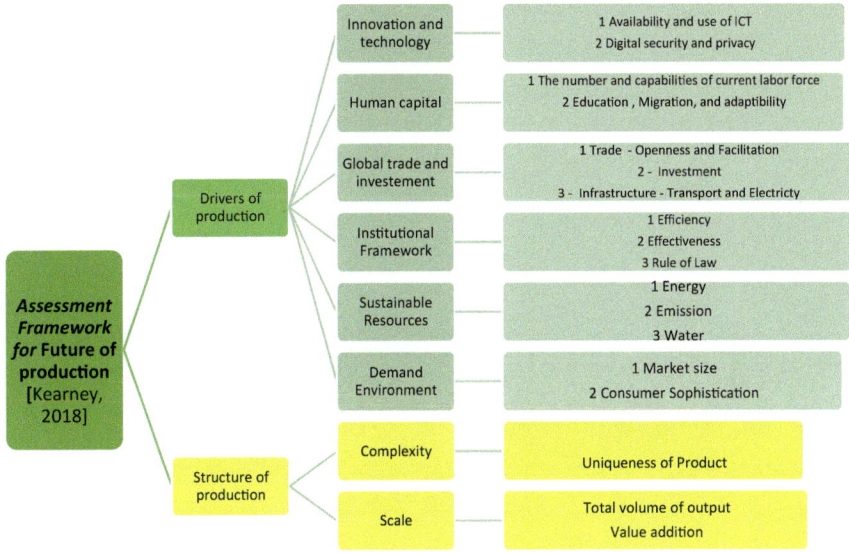

Fig. 5.10   Assessment framework for future of production (Kearney 2018)

future depending on national priority. Australia, New Zealand, Portugal, Hong, Norway, and UAE are part of this group.

- Leading—Strong current base and positioned well for future. These are a group of countries that have strong production base as well as high level of readiness for future. This group inter alia includes Japan, Germany, USA, UK, South Korea, China, and Canada.
- Nascent—Limited current base and at risk for the future. Countries in this group exhibit low level of readiness and weak parameters for drivers of production. Most of the lower and lower middle-income group countries fall in this category.
- Legacy—Strong current base but at risk for the future. This group of nations display weak performance across the drivers of production. There are ten countries including India, Russia, Hungary, Turkey, and Philippines constituting this group.

An overview of India's readiness for future of production based on ranking and valuation is provided at Box 3.

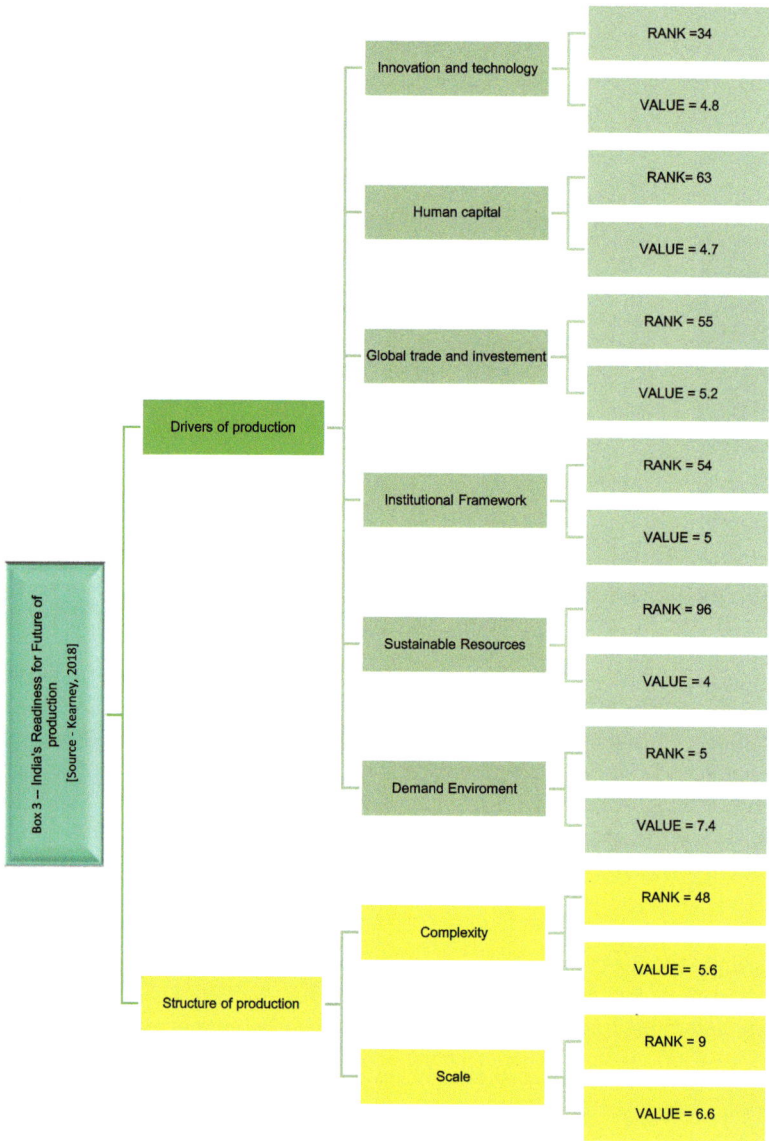

India has been placed at 30th and 44th positions in structure of production and drivers of production, respectively. Two important components of drivers of production, viz., human capital and sustainable resources are the key challenges for India (Fig. 5.11). India is way behind most of the countries in these two areas, and this clearly means that on the one hand there is an urgent need to improve educational qualification, capacity, and capability of young labor force; on the other hand, there is a pressing need to diversify energy sources in such a way that reduces GHG emissions. India is relegated to 96th position (total 100 countries) in the assessment of sustainability of resources. Figure 5.12 provides details of various components of sustainability of resources assessment for India.

India must compete with countries in the leading group and improve its ranking for both human capital and sustainability (Fig. 5.13a, b), and for this purpose an early shift to low-carbon economy will be crucial.

Fig. 5.11 Readiness assessment of India (Kearney 2018)

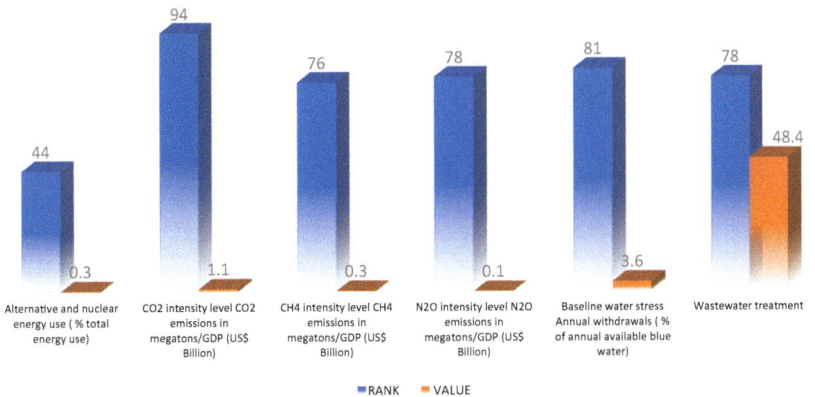

Fig. 5.12 Assessment of sustainability of resources component India (Kearney 2018)

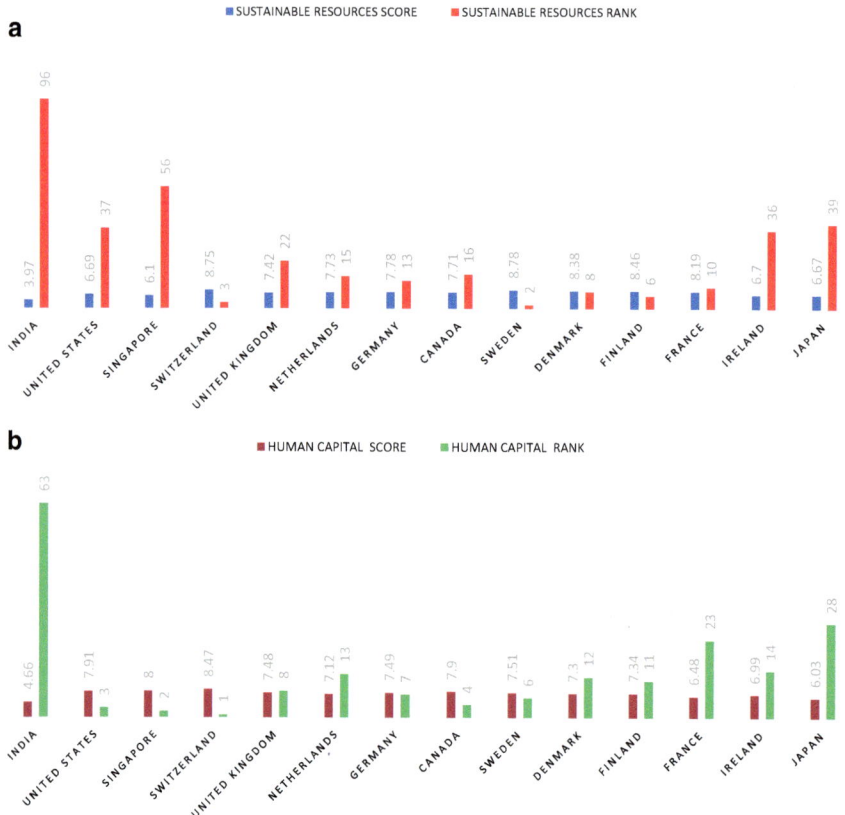

Fig. 5.13 **a** India versus Advances countries. A comparison of sustainable resources assessment (Kearney 2018). **b** India versus advanced countries. A comparison of human capital assessment (Reference—Readiness for the future of production report 2018, Published by the World economic forum, World economic forum)

Based on the overall analysis of energy situation and challenges that India faces currently and may face in future as well, an assessment of the feasibility of achieving low-carbon economy goal by 2030 is provided in Box 4.

Box 4 - Energy transformation for low-carbon economy (2030 Scenario)	
Situation	**Prospect**
Energy efficiency achieved	Possible
Shut down all thermal power plants	Not possible
Convert all sub-critical thermal power plants to super-critical	Difficult
Complete shift to renewable energy	Not possible
Low-carbon intensity (Carbon intensity = Units of CO_2 produced by a unit of energy consumed.) energy	Difficult
Cost reduction in production and transport	Possible
Remove energy-related subsidies	Difficult
Save energy consumption by increasing cost	Difficult
Provide financial incentives for renewable	Possible
Impose carbon tax	Difficult
Provide clean cooking fuel to all poor and replace wood fuel	Not possible
Optimize/reduce energy consumption with increasing income	Difficult
Increase fossil fuel prices to make renewable more competitive	Difficult
Conversion of oil-based transport sector to renewable energy	Difficult
Complete decarbonization of surface, air, and water transport	Difficult
Replace coal by natural gas for all thermal power plants	Difficult
Improve soil carbon and nitrogen content through organic farming	Possible
Improve cement quality (Cement manufacturing is one of the largest producers of carbon dioxide) to reduce frequent repair of buildings/infrastructure	Possible
Change residential and office design and technology to minimize heating/cooling requirements	Possible

One can infer from the above assessment that next three to four decades will be tough for India. With more and more climate-related disasters of increasing intensity and frequency striking various countries, India will come under severe stress of adopting low-carbon pathway, especially in rural areas. India's energy strategy for enabling a low-carbon economy should include not only the installation of new technologies for renewable energy, but due consideration must be given to a nationwide survey of use of biomass-based fuel. This will be helpful in setting the strategy and goal for the next decade or two.

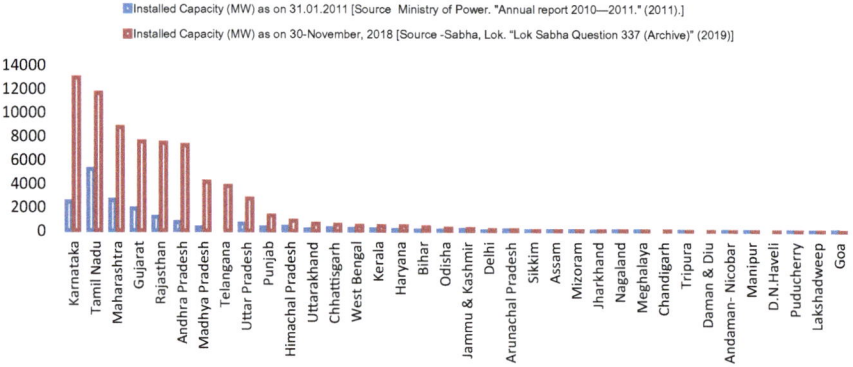

Fig. 5.14 Decadal growth of commercial renewable energy in India

5.4   Renewable Energy Assessment—India

There are six different sources of commercial renewable (1–6) and three non-commercial renewables (7–9) in India.[10] These include:

1. Wind power[11]
2. Small hydro
3. Biomass power[12]
4. Cogeneration bagasse
5. Waste to energy
6. Solar energy
7. Fuelwood
8. Agriculture waste
9. Animal waste

   The installed capacity for commercial renewables increased from 16.8 gigawatt in 2010–2011 to almost 75 gigawatt in 2018–2019 (Fig. 5.14). Some Indian states such as Karnataka, Tamil Nadu, Gujarat, Rajasthan, Andhra Pradesh, and Madhya Pradesh have progressed well while others have disappointed so far. States will have to understand that unless each one of the performs well, the dream high economic growth and competitive market will remain distant.

---

[10]*Source* Ministry of New and Renewable Energy website mnre.gov.in.

[11]The potential for wind power is assessed based on areas having wind power density greater than 200 W/m^2. The land availability for wind farm is assessed @12 ha/megawatt, all of which may not be technically feasible or economically viable for grid interactive wind power (Ref: Footnote of Table 10.27, Eleventh Five Year Plan).

[12]Biomass power ideally refers to power generated through agro-residues. But in practice biomass power generation units use fuelwood for technoeconomic reasons. A potential of 45,000 megawatts from around 20 million hectare of wastelands is assumed to yield 10 MT/ha/annum of woody biomass having 4000 kcal/kilogram with system efficiency of 30% (Ref: Footnote of Table 10.27, 11th Five-Year Plan).

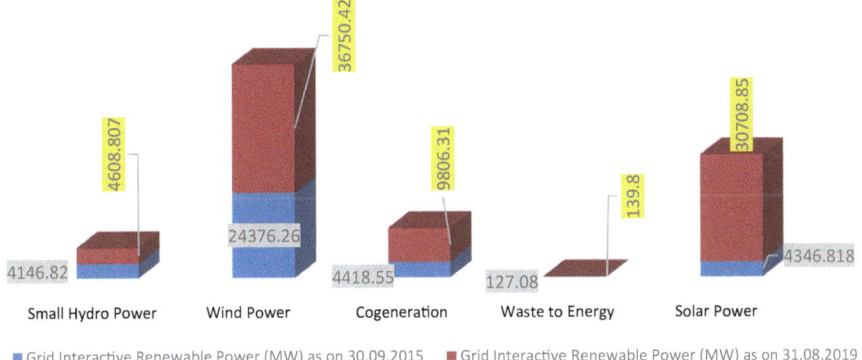

■ Grid Interactive Renewable Power (MW) as on 30.09.2015   ■ Grid Interactive Renewable Power (MW) as on 31.08.2019

Fig. 5.15 Growth of grid interactive renewable energy (*Source* compiled from ministry of new and renewable sources website https://mnre.gov.in)

From the available records, two major sources of renewable energy stand out distinctly—the wind and the solar (Fig. 5.15). Others like geothermal, biomass, and small hydro have not made much headway so far. The share of renewable energy production is expected to attain 3.12% (of total energy share) in 2021–2022 from 0.06% in 2001–2002 [Planning Commission. "12th Five Year Plan (2015)"]. This rise in two decades is, by no means, impressive considering the fact that India has the potential for wind and solar and intends to be economically stronger nation in near future. Nonetheless, expansion of renewable energy has been facing strong headwinds.

Some of the key issues for the dismal performance of renewable energy sector include:

Location specificity—Renewable energy setup is invariably location-specific and cannot be evenly distributed. For example, the states of Gujarat, Maharashtra, Karnataka, Rajasthan, Madhya Pradesh are favorable for wind and solar energy, whereas Himalayan and north-eastern states are good choice for small hydro.

Storage—Development of storage technologies has not been commensurate with the technology developments in wind and solar, due to which capacity utilization of grid connected solar and wind has been relatively poor.

Scarcity of land—Large renewable energy programs require land, which is a scarce commodity in many states. The alternative is to set up plants in wastelands and non-agriculture land which may not be viable in every situation. Eventually, one will have to go for decentralized and small-sized plants close to the consumption centers.

Transmission cost—The intermittent nature of wind and solar and variable energy production requires robust transmission infrastructure from remote locations to the load center. Not only that amalgamation of geographically dispersed energy sources would need a much larger and upgraded transmission network incurring massive investment.

	[Source -Plan. Comm., 2008]		[Source Plan. Comm., 2015]	
	Capital Cost (INR in Crore per MW)	Estimated Cost of Generation per Unit (Rs per kWh)	Capital Cost (INR in Crore per MW)	Estimated Cost of Generation per Unit (Rs per kWh)
Small hydropower	5.00 — 6.00	1.50 — 2.50	5.50 — 7.70	3.54 — 4.88
Wind power	4.00 — 5.00	2.00 — 3.00	5.75	3.73 — 5.96
Biomass power	4	2.50 — 3.50	4.0 — 4.45	5.12 — 5.83
Bagasse co-generation	3.5	2.50 — 3.00	4.2	4.61 — 5.73
Biomass gasifier	1.94	2.50 — 3.50		
Solar photovoltaic	26.5	15.00 — 20.00	10.00—13.00	10.39 —12.46
Energy from waste	2.50 —10.0	2.50 — 7.50		

Fig. 5.16 Cost of electricity generation from the renewable

Marginal impact—Renewable energy programs were taken up in 1980s (sixth plan and beyond) with the introduction of solar cooker (heavily subsidized in tribal areas @ Rs 68 apiece and Rs 400 a piece in urban areas), solar water heaters, wind mills, geothermal energy (Puga valley in Ladakh and Mani Karan in Himachal Pradesh[13]), improved cookstoves, biogas plants, and so on. Various schemes were subsequently introduced after learning lessons from the past. Nonetheless, the impact of these programs has been much below the expectations. It was initially thought that lack of institutional support at the grassroots, poor focus on training and maintenance aspects, and lack of awareness were the main impediments. Subsequently, it was also realized that continuous support to the people had raised their expectation of 'freebies' from the government and non-government organizations and therefore, their willingness to pay for acquisition and maintenance of such devices was negligible.

Generation cost—In record books, renewable energy is more expensive than non-renewable. Non-renewable energy production cost is intentionally kept low by not incorporating environmental cost and other externalities. Another important reason for higher cost of renewables is privatization. Unlike thermal, nuclear, and hydropower plants that remain under state control, much of the renewable energy sector is privatized. High initial investment and high risk discourage private sector involvement. While the cost of solar photovoltaic cell is expected to decline in future and energy will be reduced in future (Fig. 5.16), the cost of wind energy is likely to go up due to rising cost of land.

In fact, India was recognized as 'Wind Superpower'[14] way back in late 1990s because of large number of potential wind hotspots in the states of Rajasthan, Gujarat, Madhya Pradesh, Tamil Nadu, Maharashtra, Karnataka, Kerala, and Odisha. India started harnessing wind energy during 1980s by promoting wind pumps across rural areas. By 1993, the wind energy sector was opened to private investors with lucrative fiscal incentives. Consequently, wind energy development soared between 1993 and 1997 and large number of wind farms were set up. However, the investors soon

---

[13]Razdan et al. (2008).

[14]Chapter—Gone with the wind. June 30, 1999 Down to Earth.

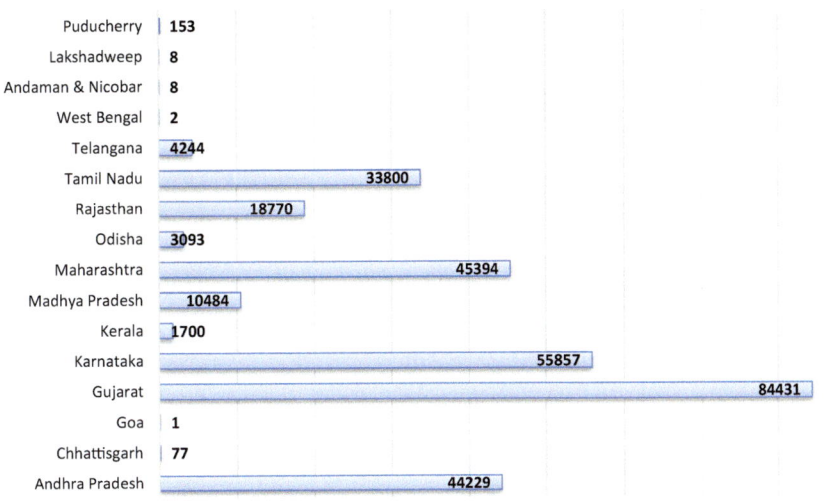

Fig. 5.17 Wind energy potential (MW) in 2017 (*Source* compiled from Ministry of New and renewable sources website https://mnre.gov.in)

realized many pitfalls, including inappropriate selection of site, grid connectivity issues, and sale and return on investment. With private sector gradual pullout, the growth of wind energy received a serious setback and by the end of eighth plan (1997) the net achievement was 860 megawatts which escalated to 5415 megawatts after a decade. In 2017, the cumulative estimated potential of wind energy (Fig. 5.17) in the country was over 300 gigawatt of which a little more than 10% was commissioned.

An ambitious target of 175 gigawatts by 2022 has been set by the government that comprises 100 gigawatts solar power, 60 gigawatts wind energy, 10 gigawatt small hydropower, and 5 gigawatts biomass-based energy. Large-scale power generation projects are in progress to achieve the ambitious target of 100 gigawatts solar power generation by 2022. At the same time, decentralized solar and other renewable energy plants of smaller capacity (2 ≤ megawatt) that can be connected directly to the existing sub-stations to save new transmission system requirement are been encouraged. For urban households this is a golden opportunity to use solar energy directly as solar cookers, solar water heaters, and solar lights or can convert it to electricity through solar panels. In order to motivate rural households, government has started a special program called 'Pradhan Mantri Kisan Urja Suraksha'. The program has three components:

1. Setting up of 10,000 megawatts decentralized ground/stilt-mounted grid-connected solar or other renewable energy-based power plants. Under this component, power plants of 500 kilowatts to 2 megawatts will be set up by individual/group of farmers or panchayats or cooperatives or other village bodies/associations.
2. Installation of 1.75 million stand-alone solar pumps each having a capacity up to 7.5 horsepower to be used for farming purposes. This program specifically intended to replace diesel pumps in off-grid areas.
3. Solarization of one million grid-connected agriculture pumps belonging to individual farmers who have grid-connected agriculture pumps.

At present, the renewable energy capacity is around 75 gigawatts (Fig. 5.18) and barring the states of Rajasthan, Gujarat, Maharashtra, Andhra Pradesh, Karnataka, and Tamil Nadu that have performed well with over 6 gigawatts capacity, all other states have lagged behind so far (Fig. 5.19). Till such time, states of Chhattisgarh, Madhya Pradesh, Uttar Pradesh, Punjab, Telangana, Odisha, West Bengal, and Haryana (Figs. 5.20 and 5.21) take proactive steps, achieving 175 gigawatts will be a distant dream.

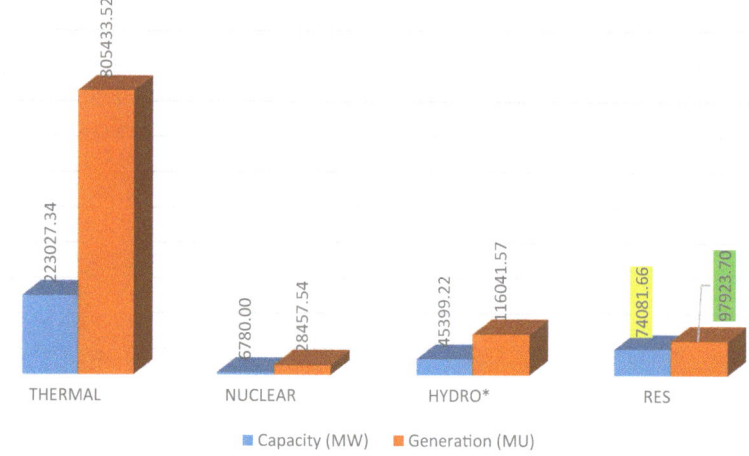

Fig. 5.18 Source wise capacity (megawatts) and Generation (Million units)in total power production during 2018–19 (up to December 2018) (Ref—*Sabha, Lok.* (*770*), 2019a)

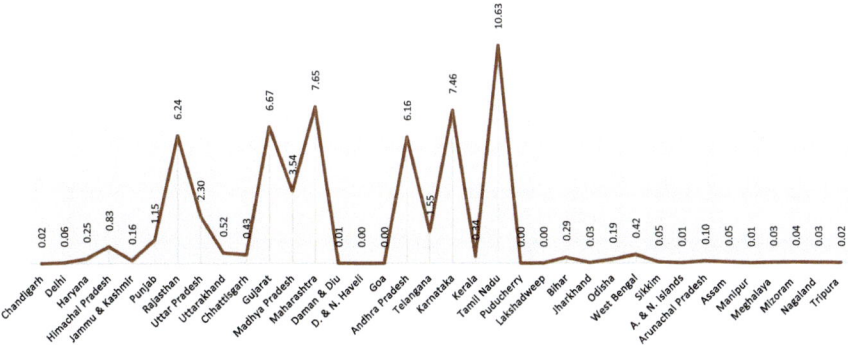

Fig. 5.19 Installed capacity of states—Renewables (GW) (2017) [*Source* Statistics, Energy 2018]

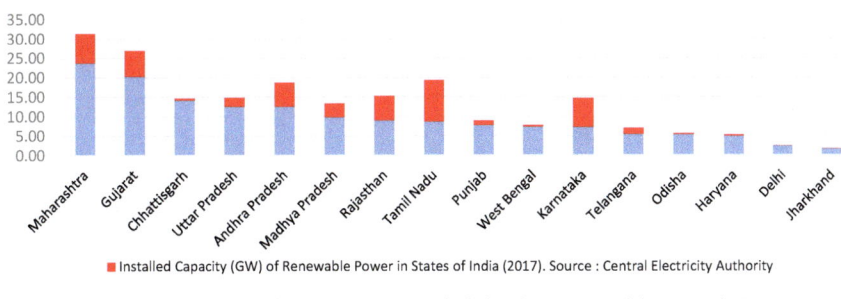

Fig. 5.20 Comparison of installed capacity (GW) thernal versus renewable in India (2017) (*Source* Statistics, Energy 2018]

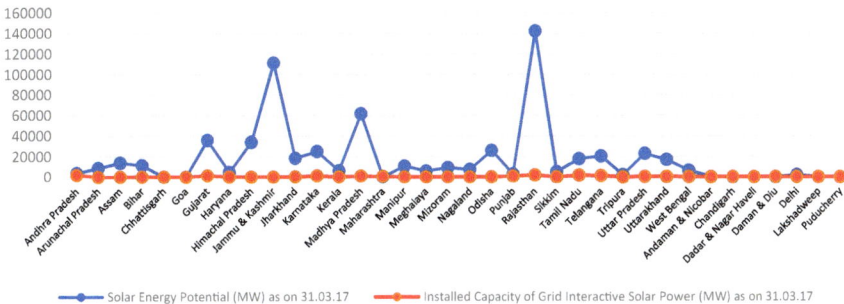

Fig. 5.21 A Comparison of solar energy potential and installed capacity *Source* Statistics energy 2018)

Box 4 - India's solar target 2022: Where do we stand

The Indian government has announced plans to enhance solar capacity to 100 GW by 2022, a five-fold increase over the previous target of 20 GW, representing a step-change in India's solar ambition. It is envisaged that around 60 GW will come from utility scale plants and the balance from rooftop and other small-scale and off-grid installations. Plans for the utility-scale installations have been taken up aggressively with the setting up of national solar mission and a series of solar parks in various states.

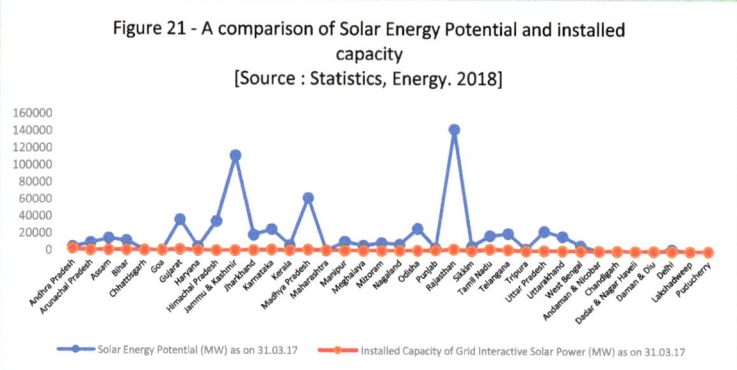

Figure 21 - A comparison of Solar Energy Potential and installed capacity
[Source : Statistics, Energy. 2018]

Nonetheless, achieving this ambitious target will face many roadblocks including land acquisition, availability of solar panel, motivation among private sector to participate, financing, and support from the state governments. There are many states that have the capacity to achieve the stipulated target (Figure 20) if they use their resources prudently and ensure sustained profitability to private players, both national as well as international. The estimated investment for solar sector would be USD 170 billion which appears to beyond the capacity of domestic sector.

[Reference – Outlook, India Energy. 2015]

Solar Cell Waste
(Ref -The Hindu (April 11,2019)

Solar cell modules are made by processing sand to make silicon. These modules are 80% glass and aluminum, and non-hazardous. Other materials used, including polymers, metals, metallic compounds and alloys, and are classified as potentially hazardous.

By 2050, India will likely stare at a pile of a new category of electronic waste, namely solar e-waste. India's PV (photovoltaic) waste volume is estimated to grow to 200,000 tons by 2030 and around 1.8 million tons by 2050. At present, very few are aware and concerned about this issue. But next couple of decades will surely witness the highpoint.

5.5 The Predicament of Biomass-Based Fuel

It is difficult to comprehend why biomass-based fuels have never been included in commercial category despite the fact that fuelwood, charcoal, and animal dung were/are sold at a decent price in villages are sub-urban areas and collection of firewood is a well-organized activity in and around forest areas. Fuelwood removal is a case of double-whammy—one, it slows down the process of carbon sequestration (and oxygen release), degrades the forest ecosystem, and reduces carbon content in the soil; and two, affects the health, income, and lifespan of women and children who are directly exposed to the smoke on regular basis. Both integrity of forests and dignity of humans are compromised, and country is subjected to ridicule in international forums and deliberations. The exact assessment regarding use of biomass-based fuel may be lacking but the following statements are a pointer to the grim reality that India has to live with.

- Tenth five-year plan, while quoting[15] NSS 55th round data (1999–2000) mentions that 86% of the rural and 20% urban households used firewood and chips or dung cake for cooking food. Besides using non-commercial fuel, rural and urban households also used LPG (5% rural and 44% urban) and kerosene (2.7% rural and 22% urban) as commercial fuels. Other primary source of cooking energy used by urban and rural household included coke and charcoal, biogas, and electricity. It is a known fact that use of traditional fuels for cooking inflicts heavy burden on women and children with increased risk of respiratory and eye infections. Lack of adequate energy supplies affect women's ability to use micro-enterprises profitably and sustainably. The NSSO survey estimated that in rural North India alone 30 billion hours were spent annually in gathering fuelwood and other traditional fuels. The economic burden of traditional biomass-based fuels, time to gather fuels, time lost in sickness, and cost of medicines was estimated to be around INR 300 billion.
- Twelfth five-year plan document acknowledged that per capita consumption of LPG in rural areas was 0.3 kilogram per month as compared to 1.8 kilogram in urban areas. This statement has two aspects. Either family size in rural India is smaller with much less requirement of fuel for cooking or people in rural areas use something else also in addition to using LPG. The latter appears more convincing to a common man who is aware of the fact that rural masses continue to use wood, agriculture, and animal waste for household needs.

---

[15](*Source* NSS, "55th Round." 2001).

- The Draft National Energy Policy (version 27.06.2017) (Aayog 2017) mentions that,

  With nearly 304 million Indians without access to electricity, and about 500 million people, still dependent on solid bio-mass for cooking, it may be acknowledged that the country has to still go a long way on securing its energy security objective.

  Energy consumption in India is characterized by low per capita level and a large disparity between urban and rural areas. In 2015–16, our per capita energy and electricity consumption at 670 kilogram and at 1075 kilowatt-hour/year, respectively, are just one third of the world average. Nearly 25% of our population today is without access to electricity and 40% without access to clean cooking fuel.

- The Economic Survey[16] report of 2018–2019 appropriately highlights the energy poverty:

  Energy poverty has been more pervasive in India than income poverty: 53% of our population could not access clean cooking in 2017 when compared to 30% for China, four per cent for Brazil and less than one per cent for Malaysia. With an increase of per capita energy consumption by 2.5 times, India will be able to increase its real per capita GDP by US$ 5000 (in 2010 prices). Additionally, if India has to reach the HDI level of 0.8, it has to increase its per capita energy consumption by four times.

A plain reading of economic survey report [Government of India. "Economic Survey 2018–19." (2019)] makes it unambiguous that every Indian must consume electricity or energy at least 2.5 times its current consumption rate. This means provision of adequate gas and electricity in all households at an enormous cost that country may not be prepared to invest in the coming decades. A snapshot of energy status and issues mentioned in the economic survey report is provided in Box 5.

---

[16]Ref—Published by the Government of India Ministry of Finance Department of Economic Affairs Economic Division North Block New Delhi-110001 E-mail: cordecdn-dea@nic.in.

Box 5

[Reference- Government of India, "Economic Survey 2018-19." (2019)]

1. India's per capita energy consumption of 0.6 tons of oil equivalent (TOE) (as compared to the global per capita average of 1.8 TOE) is much behind that of the upper-middle income countries by a considerable margin.
2. India will have to increase its per capita energy consumption by 2.5 times to increase per capita GDP by USD 5000 (at 2010 price). This will enable India's entry into upper middle-income group country.
3. India's per capita energy consumption and human development index in 2017 was 24 gigajoules and 0.64, respectively. This is required to be escalated to 100 gigajoules and 0.8 in order to facilitate India's entry into high human development country.
4. Though there has been tremendous increase in the renewable energy capacity, yet fossil fuels, especially coal, would continue to remain an important source of energy. In other words, coal-based thermal power plants may not be abandoned without completely utilizing their potential lifetime.
5. The cumulative renewable power installed capacity (excluding hydro power above 25 MW) has more than doubled from 35 GW on March 31, 2014 to 78 GW on March 31, 2019.
6. Estimates suggest that India need an investment of USD 80 billion till 2022 for renewable plants (without transmission lines) and a further investment of around US$ 250 billion would be required for the period 2023–2030.
7. If India reaches an electrical vehicle sales penetration of 30% for private cars, 70% for commercial cars, 40% for buses, and 80% for two and three wheelers by 2030, a saving of 846 million tons of net CO_2 emissions and oil savings of 474 MTOE can be achieved.

- As mentioned earlier in this chapter, the International Energy Agency[17] has estimated that 30% of India's population in 2040 (close to 480 million) mostly in rural areas will remain bereft of clean energy despite the best possible efforts from all.
- Most recent record of state-wise fuelwood use is available in report titled 'Submission on Forest Reference Levels for REDD+' submitted by India to UNFCCC[18] on January 8, 2018. Annexure 4 of the report provides state-wise annual consumption of fuelwood. As per the report, there were almost 854 million people (more than 60% population) using fuelwood of which nearly 200 million were collecting

---

[17]Ref: India Energy Outlook, World Energy Outlook Special Report. © OECD/IEA, 2015 International Energy Agency 9 rue de la Fédération 75739 Paris Cedex 15, France.
[18]Reference: FCC/TAR/2018/India dated October 29, 2018.

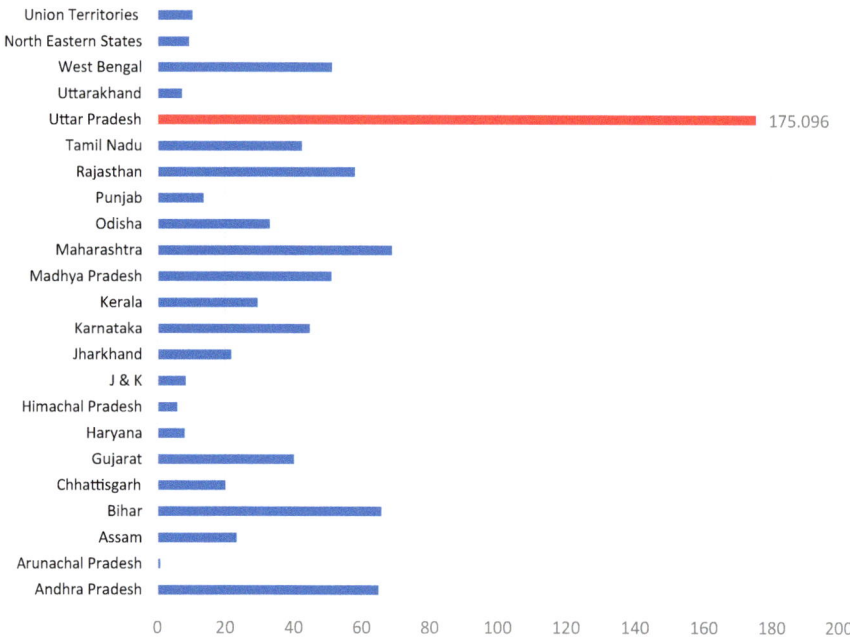

Fig. 5.22 Number of persons (million) using Fuel Wood in India (Ref-Annexure 4. Submission on Forest reference levels for REDD + submitted by India to UNFCCC on 8th January 2018)

fuelwood from forests. State-wise details are provided in Fig. 5.22 with Uttar Pradesh topping the list.

In the absence of adequate scientific data, it will be difficult to predict the actual number or range of fuelwood users, local and regional variation in quantity, and reasons for continuing use. Based on past experiences, a few reasons can be attributed to continuation of traditional cooking fuels and cookstoves and these may vary from place to place and region to region:

- Availability of cooking fuels: There are large areas in the country where wood is available aplenty either inside natural forests or outside. Roadside and community forestry plantation across the country over last four decades has increased tree density. Removal of branches of trees by local laborers is a common site across all roadsides and unprotected common lands. On the contrary, gas or kerosene have not penetrated difficult and remote areas in the country.
- Free, affordable, and non-affordable fuels: For poor daily wage earner, it is easiest to cut a few branches everyday/periodically than to opt for reasonably expensive or expensive fuel. Moreover, the traditional cookstove is almost free whereas the clean cooking stove has a cost (purchase and maintenance) attached to it.
- Aware but non-responsive: After five decades of awareness campaign, it will not be wise to accept that people are not aware of the ill effects of biomass burning on

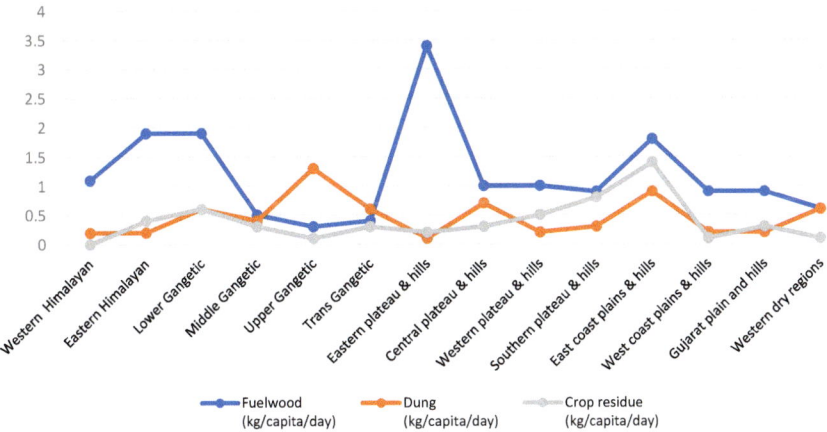

Fig. 5.23 Biomass fuel use in various agro-climatic zones in India (*Source* Ravindranath and Hall 1995)

their health. At times, it is difficult to figure out that biomass users possess mobile phones but at the same time look for freebies from government in the form of clean cookstoves and LPG. There seems to be no justification except the mindset of many such individuals who have vowed to indulge in free supply till it lasts.

- Design of cookstove and appliance: For various reasons including size and shape of cooking utensils, the shape and size of cookstove will vary. In many states people rejected several models of smokeless cookstove because their utensils were either too big or too small.
- Availability of commercial fuels: Even in those areas where LPG has been distributed, people invariably keep stock of fuelwood as backup as they are unsure of LPG supply as well as their future income and savings.
- Biomass use per capita varies in different agroclimatic zones (Fig. 5.23) depending on total energy requirement as well as availability of fuel. But there is no nation-wide authentic quantitative and qualitative data on the impact of removal of biomass on soil, forest and agriculture productivity, decline and loss of species, and human health during the last 100 years. And the cumulative cost will be astronomical in terms of climate change, health, and economic impacts.

Let us examine four different scenarios with the presumption that each individual uses only 1 kilogram of fuelwood each day. The four scenarios use dried animal dung, crop waste, and coal as replacement for fuelwood with the following conversion factor:

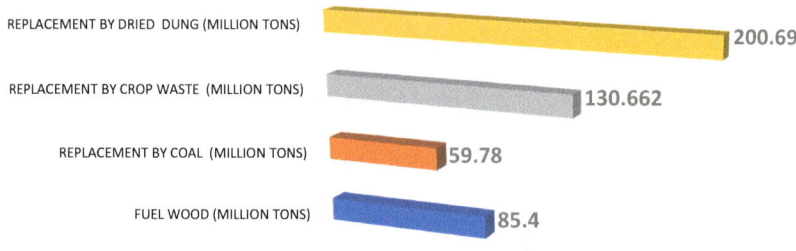

Fig. 5.24 Scenario 1 (854 million use fuelwood for 100 days a year). Ref—India submission on Forest reference levels for REDD+

| 1 kg of fuel wood = 0.7 kg of coal |
| 1 kg of fuel wood = 2.35 kg of dried dung |
| 1 kg fuelwood = 1.53 kg crop-waste |

Scenario 1: 854 million people use fuelwood for 100 days and other fuels for the balance period (Fig. 5.24). Under this scenario India will need an additional 60 million tons of coal or 130 million tons of additional crop waste or 200 million tons of additional dried animal dung if fuelwood use is completely stopped. If fuelwood consumption continues as such, at least 145 million tons of CO_2 will be added to the atmosphere every year.[19]

Scenario 2: 854 million people use fuelwood for 200 days and other fuels for the balance period (Fig. 5.25). Under this scenario India will need an additional 120 million tons of coal or 160 million tons of additional crop waste or 400 million tons of additional dried animal dung. If fuelwood consumption continues as such, at least 290 million tons of CO_2 will be added to the atmosphere every year.

Scenario 3: 854 million people use fuelwood for 300 days and other fuels for the balance period (Fig. 5.26). Under this scenario India will need an additional 180 million tons of coal, 392 million tons of additional crop waste, and 600 million tons of additional dried animal dung. If fuelwood consumption continues as such, at least 435 million tons of CO_2 will be added to the atmosphere every year.

Scenario 4: 854 million people use fuelwood for 365 days and other fuels for the balance period (Fig. 5.27). Under this scenario India will need an additional 218 million tons of coal, 477 million tons of additional crop waste, and 732 million tons of additional dried animal dung. If fuelwood consumption continues as such, at least 530 million tons of CO_2 will be added to the atmosphere every year.

One can work out a series of different scenarios using different permutations with different types of fuels, renewable or non-renewable, commercial, or free. What is

[19]1 kilogram of wood when burnt releases 1.7 kilogram of CO_2 on average.

Fig. 5.25 Scenario 2 (854 million use fuelwood for 200 days a year). Ref—India submission on Forest reference levels for REDD+

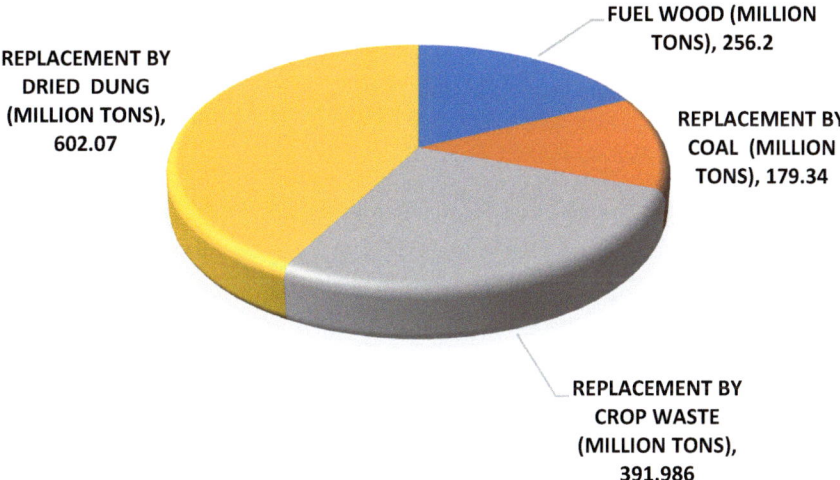

Fig. 5.26 Scenario 3 (854 million use fuelwood for 300 days a year). Ref—India submission on Forest reference levels for REDD+

important for all of us is to understand that use of biomass is extensive across India and will continue for several decades if corrective measures are ignored. Not only that, over the last several decades, the naturalness, authenticity, and integrity of our natural forests has been compromised severely due to slow poisoning by fuelwood collectors. Non-commercial energy sector has been ignored for seven decades both from scientific and economic perspectives. It is difficult to comprehend as to why economists have failed to assign monetary value to fuelwood, animal dung, and agriculture waste. This approach has done incalculable harm to the country that is several times the cost of LPG that has been found to be a viable substitute (Fig. 5.28).

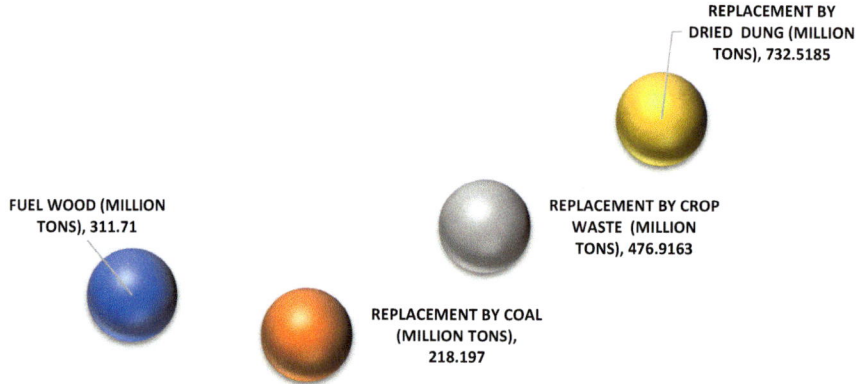

Fig. 5.27 Scenario 4 (854 million use fuelwood for 365 days a year). Ref—India submission on Forest reference levels for REDD+

   Fuelwood production data for the decade 2006–2016 provided by Food and Agriculture Organization (Fig. 5.29) shows almost no change in the production of coniferous and non-coniferous fuelwood. Assuming an average quantum of 300 million cubic meter per annum, a conversion factor of one cubic meter of fuelwood to 500 kilograms, and an average consumption of 1 kilogram per capita per day, one can safely conclude that more than 410 million people were using fuelwood only to say the least. Also, for producing 300 million cubic meters of fuelwood, a minimum of 300 million hectare of land (including forest land[20]) with an average productivity of one cubic meter per annum would be required. In essence, we have and will continue to pay heavy price by neither providing substitute nor commercializing biomass-based fuels.

   India has lost tree biomass much in excess of its productivity potential and sustained yield principle. And this process will continue till the time fuelwood users switch over to alternatives. The ongoing trend indicates that an assured supply of alternative energy substitute for all biomass-dependent families may take several decades. Under these circumstances, heavy investment would be required to raise tree plantation (for future supply and as a compensation for past deforestation and soil degradation). Average biological productivity of Indian forest is appallingly low as compared to world average,[21] and therefore, it is extremely important to restore the health of natural forest and conserve biological species in situ lest we reach the point of no return. Realistically, two major programs are to be launched in a big way. One, restoration of natural forests and two, energy plantation (outside natural forests) to ensure sustainable supply of fuelwood to people till the time fuelwood consumption is reduced to naught. Both are required to be done in a highly scientific manner adhering to the precautionary principles.

---

[20] Average productivity of forest land in India is close to 0.5 cubic meter per year.

[21] As per FAO—Global Forest Assessment Resources 2005, average global forest productivity is—116 CMT per hectare per year. Indian forest productivity is 0.5 CMT/ha/year.

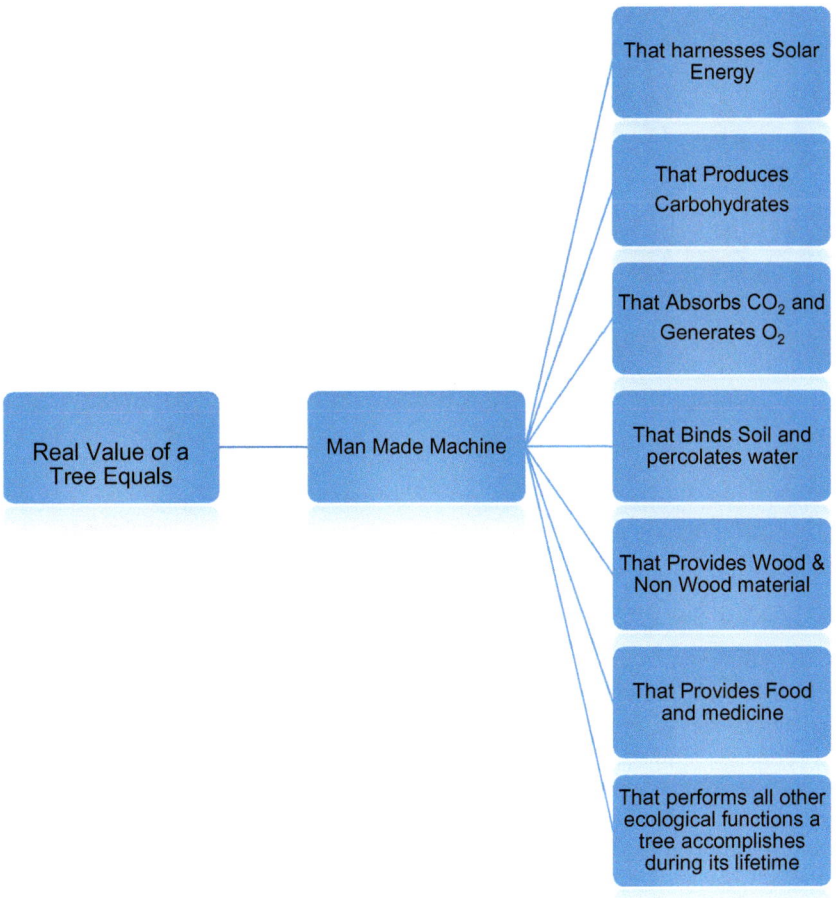

Fig. 5.28 Real value of a tree equals

We must also remember that services provided by the energy are more significant than the quantum of energy, and therefore, energy performance will overshadow energy supply or energy use. As far as fuelwood is concerned, this can be achieved in the following science-based approach (Srivastav 1992), which is as follows:

• Land for energy plantation can be ascertained from the following formula:

$$\text{Area(Hectare)} = \frac{\text{Number of families} \times \text{Average annual consumption of wood fuel per family}}{\text{Annual volume increment in the fuelwoodlot}}$$

• Species selection for each area can be done by ranking different local tree species with the help of following formula:

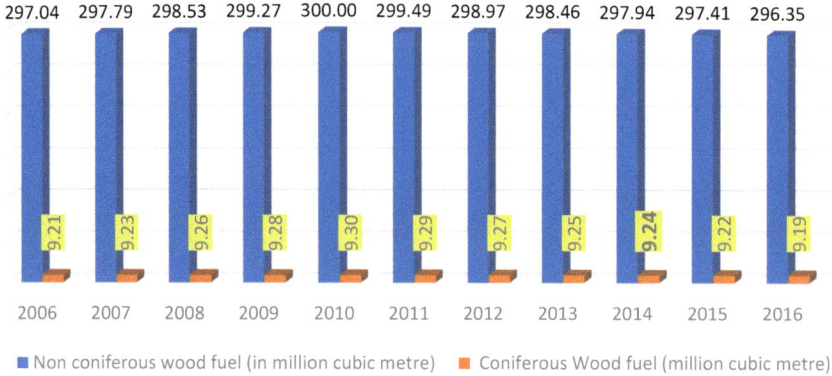

Fig. 5.29 Production of fuelwood in India (MCM) (Ref Faostat 2018)

$$\text{Species rating index} = \frac{\text{Wood density} \times \text{Diameter of tree at breast height(meter)} \times \text{Tree height(meter)}}{\text{Fuel value(kilogram)} \times \text{Age of tree at harvest}}$$

Fuel value[22] of different tree species will depend upon factors such as type of cook stoves, thermal conductivity of appliances, quality of water, seasonal temperature fluctuations, and type of food. Species with high rating index can be promoted for fuelwood lots in the vicinity of human habitation. The consumption of wood can be brought down by promoting the use of energy-efficient cookstoves and utensils with high thermal conductivity. The ashes collected from the hearth can be utilized for improving the soil productivity of plantations.

5.6 Every Little Step Will Be a Great Leap for India

Rural India will be the key to India's success during fourth industrial revolution. Health, education, skill development, and clean energy security and access in the coming decade or two will decide the future of this country. Per capita rural energy consumption has to be brought close to that of urban consumer. With assured and uninterrupted supply of electricity, individuals will gradually switch over to contemporary ways of cooking over a period of time in the same way as transport system has changed. Villagers have to be given basket of cooking fuels and flexibility in fuel usage with sustained supply of fuel, including LPG, biogas, solar cookers and

---

[22]Fuel value can be calculated as follows: For example, if 5000 kcal are required to cook 1 kilogram of rice under **ideal** conditions, but in practice, if 2 kilograms of wood of *Prosopis cineraria* (Calorific value = 5000 kcal/kilogram) is consumed for cooking 1 kilogram rice, then the fuel value of *P. cineraria* in that locality will be 2 kilograms.

heaters, fuelwood, charcoal, which would ensure security and access, provided that the cost is based on the calorific value and not market price. People have to be gradually dissuaded from using fuelwood and animal dung. Local entrepreneurs should be involved in this process product design, engineering, manufacturing, marketing, maintenance distribution, and sales.

Each unit of electricity that people consume globally releases 0.19 kilogram of carbon di oxide. This is a conservative figure and actual release may vary with the type of source and technology used for electricity generation. In case of India this figure may be close to 1 kilogram of carbon dioxide. In comparison, release of carbon dioxide will be insignificant if all thermal power stations are theoretically converted into solar/wind/nuclear/hydropower or a combination of these. After examining the past energy trend in India, one can safely presume that full decarbonization is a difficult option for next three decades if not more. An average Indian upper-middle and rich Indian home generates between 2 and 3 tons of emission (Fig. 5.30), and therefore, improved insulation of their houses, redesigning internal structure for central heating/cooling, installing double-glazed windows for better cooling/heating, use of solar water heaters for warming house as well, and judicious use of electrical appliances will reduce emission in a big way. Besides personal dwellings, it is a common experience that most of the public offices, corporates, shopping complexes, educational institution, and hotels in India overuse electricity by excessive lighting, not switching off when energy is not required, and so on. Appropriate action by all stakeholders will reduce CO_2 emission by billions of tons and lower atmospheric temperature as well.

Another good example of carbon dioxide release is use of hot water during winters (Goodall 2010). Presuming an average of 20 L of hot water per individual per day, let us see how much energy is required:

i.   Hot water required per person per day = 20 L.
ii. Energy needed to heat water by 1 °C = 0.0011 kilowatt-hour.
iii. Total energy required to heat 20 L by 50 °C (say from 10 to 60 °C) = 0.0011 × 50 × 20 = 1.1 kilowatt-hour or 1.1 unit.

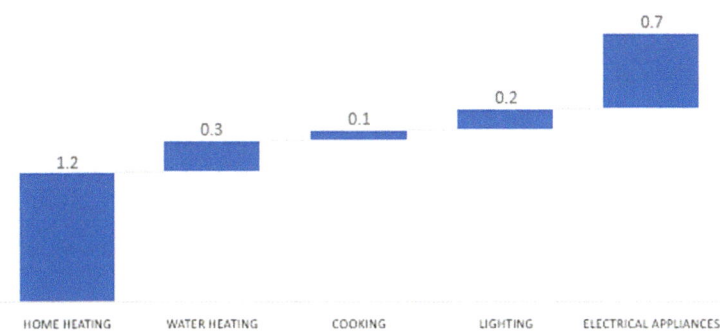

Fig. 5.30 Average emissions from home (Tonnes) (*Source* Goodall 2010)

iv.   Total energy required for 4 persons (an average Indian family size) $= 1.1 \times 4$
      $= 4.5$ kilowatt-hour.
v.   Total energy required by a family for four months $= 4.5 \times 120 = 540$ kilowatt-
      hour.
vi.   Total carbon dioxide released by a family $= 540 \times 0.19 = 102.6$ kilogram.

102.6 kilogram carbon di oxide is a conservative estimate as more than 60%
population of India resides in rural areas and uses biomass-based fuel that will escalate CO_2 release exponentially. Instead of using electricity or wood, if majority of
us decide to use solar water heaters then the emission level can be brought down
considerably. This is just one instance of lowering emission level. Adopting energy-
saving technologies and products and changing consumption behavior and lifestyle
will go a long way in not only reducing emission but reducing cost of energy as well.

One of India's remarkable energy-saving accomplishment has been shift from
incandescent to solar lights. Incandescent bulbs, originated in early nineteenth
century, were lifeline of every household almost till the end last century. Though
inefficient (in current context) and a source of heat radiation, they are cheap and still
popular in rural India. Sometimes during middle of the twentieth century fluorescent tube lights, which used mercury, were introduced that continue to be popular
even today. During the decade of 1990s, halogen bulbs were introduced, which were
more expensive, lasted longer, and gave slightly more illumination per unit of electricity consumed than the conventional incandescent bulb. But these bulbs generated
more heat all of which was wasted. Not highly successful commercially, the halogen
and incandescent bulbs were replaced by compact fluorescent lamps (CFL). Though
more expensive than their contemporaries, the CFL lasted at least ten times longer and
generated four to five times more illumination per unit of electricity. Currently light
emitting diode (LED) bulbs are most popular in India, both rural and urban and one
of the reasons is that the cost of low watt LED has been kept exceptionally low. These
bulbs can last up to 100,000 hours and consume 80% less electricity than conventional incandescent bulbs. In fact, LED bulbs and strip lights have brought down
electricity consumption substantially. A comparison of different lighting devices is
given in Fig. 5.31.

In the same way, emission from transport sector particularly cars and two wheelers
are worse than home emission and varies with the number, average distance trav-
eled per day, type of engine, and quality of fuel used. Public transport and electric
vehicles are the best options under the circumstances (Fig. 5.32) but that may not be
acceptable as vehicle manufacturing and ancillary units employ millions of Indians.
Nonetheless, several Indian cities have introduced mass transit system with great
success. In the short term, three steps are required to curtail emission—limit the
number of personal automobiles to the carrying capacity of each city, incentivize
commuters using electric vehicles and bicycles, and reduce air travel by introducing
high-speed trains in such a manner that total time taken (origin to destination) in
commuting is almost equal.

Some other steps taken by India to overcome energy challenges (Chengappa 2015)
are as follows:

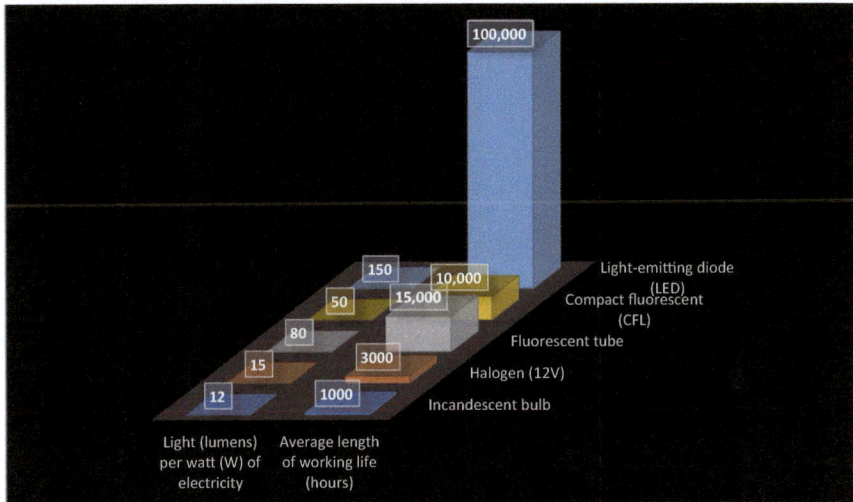

Fig. 5.31 A comparison of different types of lighting bulbs (*Source* Goodall 2010)

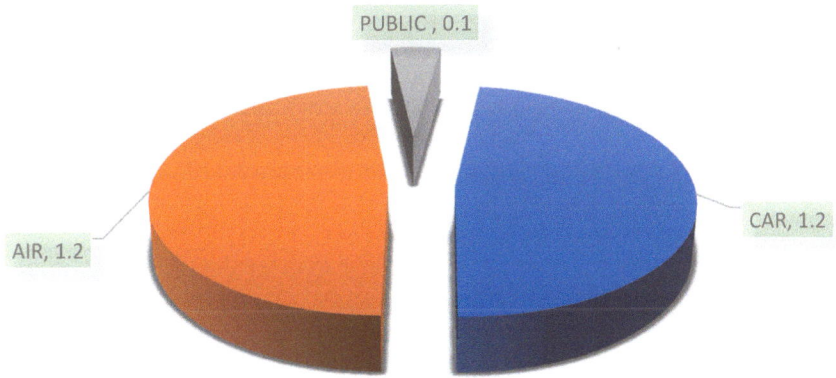

Fig. 5.32 Average emissions from transport (Tonnes) (*Source* Goodall 2010)

1. Smart grid—This is basically an energy-conserving measure that prevents wastage and improves efficiency by tracking electricity distribution and use in real time. With two-way digital communications between power plants and users, smart grids can balance supply and demand in real time and make consumers active participants in the production and consumption of electricity. As the share of generation from variable renewable resources such as wind and solar increases, a smart grid can easily handle fluctuations. This innovative and automated system can change the way electricity is distributed and consumed through the use of smart meters, smart appliances, and renewable energy. For India, this will be key to success of such initiatives and 24 × 7 power supply to all citizens, developing

100 smart cities, 175 gigawatts renewable energy, as well as 6–7 million electrical vehicles.

2. LED revolution—Light emitting diode bulb, first manufactured in India in 2009 to replace incandescent and CFL bulbs, has several advantages. It generates less heat and contains no mercury, has more longevity, and are eight times more energy-efficient than incandescent and twice that of CFL. Introduction of LED bulbs is expected to save around 16 billion tons of CO_2 release in next 25 years.

3. Energy-efficient air conditioners—On average, air conditioners consume a third of power consumption in Indian households due to repeated cycles of compressing gas to liquid for cooling hot outside air. Initially, CFCs were used as refrigerant, but they were found to be depleting ozone in the atmosphere. The CFCs have been replaced with halogenated chlorofluorocarbons and hydrofluorocarbons. These new refrigerants cause faster cooling and can save energy up to 20%.

4. Biodiesel—This is one of the most efficient alternative fuels and emits 78.5% less CO_2 in comparison with petroleum diesel. There are more than a dozen companies in India producing biodiesel and blending of petroleum diesel with biodiesel is allowed by the government.

5. Hybrid cars—These are fuel-efficient and eco-friendly vehicles that use energy from two different sources—an electric motor and petrol/diesel engine. These cars enhance fuel efficiency by 30% and reduce GHG emission by 30%. India's plan is to have six million cars on road by 2020.

6. Advanced ultra-super-critical technology for thermal power plants—This is a state-of-the-art technology that generates 30–40% more energy from the same amount of coal through high temperature and pressure. A 4000 megawatts advanced ultra-mega power project can save 4000 million tons of coal per year. India intends to set up such plants in near future.

7. Wind turbines—These convert kinetic energy from wind into electrical energy and is the cleanest source of renewable energy. The blades of wind turbines, made of steel and glass, make 10–22 revolutions per minute to generate electricity. Efforts are on to have light-weight blades of carbon fiber to increase its revolutions.

8. Concentrated solar thermal power—Solar power has two major technologies: solar photovoltaic systems and concentrated solar power. Solar photovoltaic systems convert solar energy directly into electricity. Concentrated solar power uses mirrors to concentrate sunlight and generates steam to drive turbine for generating electricity. A collaborative project at Mount Abu has been set up by World Renewable Trust and Ministry of New and Renewable energy with German Collaboration to meet the energy requirement of 20,000 people.

5.7 Fourth Industrial Revolution and Global Energy Prospects

External energy obtained from fossil fuel has pervaded modern-day life of most of us, minimizing the use of internal energy that we all possess. From human-driven ploughs, cart, bi, and tricycles to animal-driven carriages, carts, and tills, technology has reached its zenith where cars are driven on voice command, computers, and many other electrical and electronic devices are operated effortlessly through interactive voice response system. The third wave of invention set off by advances in computing, information and communication technology uses external energy that has a cost. Coal is dirtiest adding substantially to carbon emission, crude oil is subject to shocking fluctuations in output and price, natural gas supply has been monopolized by few countries, hydropower faces displacement outcry, and nuclear power has a flip side of waste disposal and radiation leaks. Impact and sensitivity to dirty fuels has motivated technical experts to look for three options. One, how to minimize the bad effects of dirty fuels; two, look for cleaner sources of energy; and three, introduce low-energy consumption devices. Super-critical technology for thermal plants, small hydropower plants, and LED bulbs are good examples to exemplify these. The transition for cleaner energy will take time given that hydrocarbons in earth's crust are still aplenty, the extraction technology is yet to 'peak', and shift from machines operating on dirty fuels to cleaner fuels will depend on investment, production cost, and purchasing power. Investment in cleaner fuel is inversely proportional to the cost of dirty fuel, and the cost of dirty fuels is decided by conglomerates that ensure buoyancy. Fortunately, public opinion and Paris Agreement has benefitted cleaner energy with unprecedented investment in renewables[23] by rich nations. But this has not deterred them to invest less in dirty fuel. On the contrary, USA and Canada have invested huge money in shale oil and gas and tar sand oil. Russia has invested in natural gas extractions and Germany has not reduced its coal mining.

The fourth industrial revolution is expected to change the following:

i.   Shift in production paradigm—from mechanization and human skills the fourth revolution will witness discrete role of artificial intelligence, robotics, internet taking human skills to next level of mental excellence.

ii. Improve the quality of life and income—by improved transportation (e.g. electric vehicle, mass transit system), ease of driving, less pollution, and so on.

iii. Alteration in labor demand pattern—more automation means less intervention by humans. For example, driverless cars, robotic, and vacuum cleaners have changed the demand for labor. This will also mean better adaptiveness by countries with right mix of human capital. For countries like India with large workforce, it will be important to invest money in training and retraining people to suit market requirement both nationally and globally.

---

[23]The average investment in renewables was around USD 260 billion per year during the first half of last decade. Ref Special report on Energy and Technology, The Economist, January 17, 2015.

iv. Innovators and intellectuals will be in huge demand—countries with storehouses of intellectuals and innovators will be the wealthiest vis-à-vis country having unskilled and semi-skilled manpower.

v. Shift in energy production—It is expected that the fourth industrial revolution will be less damaging to the environment and usher in more sustainable production systems. Production systems today are responsible for 35% of all global electricity use, generate 20% of CO_2 emissions, and account for a quarter of all extractions of primary resources adversely impacting the environment by the over-exploitation of natural resources, the pollution/destruction of ecosystems, and reduction in biodiversity.

The task before India is arduous for three reasons. The population growth, lack of adequate skill among the young and middle age group, and sustainability of resources. India will have to continuously endeavor for diversifying its energy resources besides developing a cadre of highly skilled (especially digital skill) manpower to counter the challenges from competing nations. Time moves at an unprecedented pace and next three decades will be extremely challenging for India as the fourth phase of industrial revolution is all set to change the global energy landscape.

The roadmap for acquisition and acceptance of cleaner energy is beset with potholes. Since wind does not always blow and sun does not shine all the time in all parts of the globe, it is difficult to match electricity output with demand all the time. While many developed countries, including India are fast moving toward solar and wind energy, but their intermittent supply makes efficiency and storage of energy equally important. Scientists are now working to develop batteries that can store huge amount of electricity and make it available to grid operators when required. Efforts are ongoing to have smaller power stations well distributed over large area rather than a large power station with grid connectivity spread all over. This makes lot of sense for renewables as one can generate and store electricity cheaply and reliably if these smaller units are combined with better quality light-weight batteries with vast storage capacity. In short, the renewable energy sector is aspiring for the following:

• Maximize conversion of solar and wind energy to electrical energy per unit time and area.
• Develop batteries that are small, compact, and lightweight with very high storage capacity.
• Minimize per unit cost of electricity so generated to less than that generated from non-renewable sources.

Researchers in many countries are currently looking for solar cells that can absorb maximum photons per unit area and store maximum energy in small and light-weight batteries to ensure sustained supply of energy during lean period. Solar cells are made of semiconductors such as silicon and perovskites.[24] Traditional solar cells based on silicon semiconductors are not very effective in converting photons to electricity. The reason is that silicon semiconductors absorb energy from the photons of shorter

---

[24] Perovskite is a calcium titanium oxide mineral composed of calcium titanate ($CaTiO_3$). Perovskite solar cells convert ultraviolet and visible light into electricity very efficiently.

wavelengths only and the energy from long wavelengths is lost. But there are other elements such as gallium and indium that can absorb solar energy in the infrared and ultraviolet spectrum. Improved version of solar cells consists of a stack of four, one on top of each other where each layer in the stack is made of different materials such as gallium and indium and stacked in such a manner that each layer absorbs energy from certain part of the spectrum, converts it efficiently into electrical energy, and passes the rest onto the next layer. Future solar cells will be made of substances that will absorb sunlight in invisible spectrum and allow visible light to pass through. This will bring in revolution in solar energy as we will then have solar energy produced from wind shields of cars, buses, glass panes of buildings, and so on.

Besides silicon, solar cells are also made of perovskites that are more efficient. These are compounds that share similar crystal structure and are named after the mineral that was first found to have this structure. These compounds are cheaper than silicon and can convert almost 20% of the sunlight falling on it into electricity.[25] One major difference between these two semiconductors is that in a silicon-based solar cell, a thin wafer of 200 micron is used that is sliced off from the bigger block. However, a perovskite solar cell can be made by dispensing the chemical solution on a surface and making a cell of desired thickness. Moreover, perovskites can be made with different combinations that can absorb maximum photons in different parts of the visible spectrum. Each type of semiconductor has a property called band gap that determines the longest wavelength of light a semiconductor can absorb as well as maximum amount of energy that can be captured from photons of shorter wavelengths. The photons absorbed are converted into electrons which are gathered by electrodes to flow into a circuit.

The fourth industrial revolution has the potential to change the energy outlook of the world. We have already graduated from wood–coal–oil era and entered white gold stage that is dominated by nanotechnology. With its unique physical, chemical, mechanical, and optical characteristics, this technology offers solution by squeezing the size of large heavy-weight batteries (such as lead-acid) using new class of material called nanoparticles. The fundamental premise of nanotechnology is that all materials are made of atoms and possess distinct chemical properties that depend on the structure of the clouds of electrons. Sometimes an atom will pair off one of its electrons with an electron from a neighboring atom to form a chemical bond and form a molecule or a kind of crystalline structure such as semiconductors. Advocates of nanotechnology aim at building things atom by atom so that we can have floods of new materials and new inventions.

One of the significant applications of nanotechnology has been in lithium batteries. A typical lithium cell consists of two electrodes (a cathode and an anode) where anode is made of graphite and cathode is of lithium cobalt oxide. The electrolyte used in the cell is made of a solution of lithium salts and organic solvents. During the charging of cell, positively charged lithium ions in the electrolyte move toward negatively charged anode and gets deposited. When the cell is used for operating any device, the electrons flow from the anode into the device circuit and re-enter the cell

---

[25]Science and Technology, The Economist, May 16, 2015.

via cathode. On average, a lithium battery can store 100–250 watt-hours/kilogram which is more than twice of what nickel-cadmium can do. A small electric car with 25 kilowatt-hour lithium battery can cover 175 km before recharge (Johnson 2009). Research and technological inputs since 1991, when lithium batteries were commercially introduced, have tremendously improved the energy density of lithium batteries by developing thin films (single-atom thick) to enclose positive electrode coupled with a negative electrode of sulfur (like lithium, sulfur also has a very high energy capacity). This has enabled a battery to hold about five times as much energy by weight as compared to previous lithium batteries. Besides, car, lithium batteries are being used for storing power generated through solar panels and windmills in the USA, Australia, and South Africa.

Lithium, which was initially used to treat bipolar disorders, ceramic, and nuclear weapon industry, has now emerged as the most essential component of all digital devices. It is the lightest metal in the periodic table, heat-resistant, and is highly reactive with three electrons in its atom of which two are tightly bound to its nucleus but the third can be easily dislodged to form positively charged lithium ion. This makes lithium inherently unstable, and therefore, instead of using lithium in its metallic form experts opted for safer compounds containing lithium ions. The storage ability of a lithium battery depends on its energy density, that is, the amount of energy that can be stored for a given weight or volume. A typical lithium battery can store 100–250 watt-hours/kilogram[26] and an electric car with 24 kilowatt-hour lithium battery has an average range of 175 km. Improvement in energy density by using a combination of manganese, nickel, cobalt, and graphite electrodes is being experimented with. These and other properties make lithium more efficient and valuable as compared to heavier batteries of the past made of lead, zinc, and nickel-cadmium.

These extraordinary properties attracted many in 1970s to explore the possibility of lithium battery as an alternative to petroleum. It was found that a battery with lithium anode and titanium di-sulfide cathode worked well in providing electricity and was rechargeable as well. In fact, Sony was the first company to use lithium ion battery in consumer electronics in 1991. Subsequently, in 2004, another semiconductor called graphene was discovered that is extremely thin (single-atom thick and can be used as a two-dimensional material), light, strong, and transparent. While it gave hope to the electronic industry that was looking for small-sized rechargeable cells, this device was not accepted as lithium was found to be highly inflammable. During its life, a thin layer of lithium is deposited on the surface of electrodes which is sloughed off by the continuous contraction and expansion of electrodes and is replaced by another layer of lithium. This process eventually saps the battery of its lithium ions. Being highly reactive element, overcharging, or manufacturing defects in batteries make them prone to short circuiting, heating, and explosion. Subsequent researches improved lithium ion batteries and revolutionized the electronic world. The most common example is that of battery-operated torches that worked on small incandescent bulb and disposable batteries. These have been largely replaced by

---

[26] A nickel cadmium battery can store only half of this.

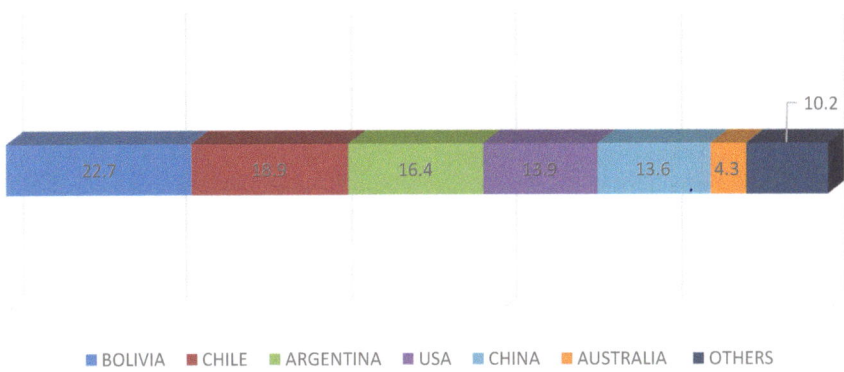

Fig. 5.33 World's indentified Lithium sources (% of total Lithium—40 million tons) Ref the economist, January 16, 2016

mobile phone torches that use lithium batteries. Eventually all such devices that use disposable batteries will be replaced by an integrated and compact device using lithium batteries.

Lithium is known to be present in every continent except Antarctica, but the largest reserves are concentrated in Bolivia, Chile and Argentina, also known as the 'Lithium Triangle'. Lithium reserves around the world are estimated at 53 million metric tons with Bolivia in the lead. Almost half of today's lithium is mined for battery-related purposes only. With the steep increase in battery demand, lithium supply is expected to increase by a factor of 6 from 2018 to 2030. With limited availability (Fig. 5.33), the price[27] of lithium carbonate has risen from 4000 USD per ton in 2008 to almost 14,000 USD in 2015. Lithium batteries are relatively expensive costing USD 500/kilowatt-hour capacity.[28] This means that a battery pack for a small electric car would cost around 10,000 US dollars, an expensive proposition for a common man to switch over to the new cars.

5.8 Great Hope for Future—Fusion Power and Fuel Cell

1. *Fusion Power*

The simple equation $E = mc^2$ derived by Albert Einstein can provide incredible amount of energy and solve the problem of energy drought once and for all. The equation implies that energy (E) of a substance is the product of its mass (m) and the square of speed of light (c^2). Since the speed of light is extremely high, a tiny mass can, when completely converted, result in incredible amount of energy. This

---

[27]Ref—The Economist, January 16, 2016.

[28]Ref—The Economist. May 30—June 5, 2015.

kind of conversion is possible in nuclear reactions where basis changes are initiated in the nucleus of atom. Two known types of nuclear reactions where mass to energy conversion occurs in a remarkable way are fission (splitting of the nucleus of an atom) and fusion (combining of nuclei of two atoms).

The solar energy, which we are trying to convert into electrical energy, is essentially a product of fusion reaction that goes on incessantly inside the sun, whereby nuclei of hydrogen atoms continuously fuse to form nuclei of helium atoms releasing tremendous amount of energy in the process. But unlike fission energy that has been produced on commercial scale, power from fusion energy is still many years away. One of the major impediments in this process is that it is very difficult to fuse two atomic nuclei as both are positively charged and thus repel each other. What is required is to give sufficiently high velocity to these nuclei by heating them to hundreds of millions of degrees as happens in case of Sun. Once fusion starts, it will generate enough heat for other atoms to fuse and start a chain reaction. That is why fusion reaction is also called thermo nuclear reaction or nuclear reaction caused by heat.

Energy from human fusion process, whenever realized, will change the landscape of this planet with sustainable solution to the climate-related threats. There will be zero GHG emission and no long-lasting radioactive waste disposal issues. Scientists have been relentlessly working for more than seven decades to realize the dream of human controlled nuclear fusion that will be carbon-free and thus change the energy landscape of the planet. There are three primary conditions that need to be fulfilled in a fusion reactor—high temperature exceeding 100 million °C for collision of hydrogen atoms, optimum plasma density to enhance probability of collision, and sufficient confinement to hold the plasma and ensure continuous collision. There are many types of fusion reactors being tested in various countries such as 'Tokamak', 'Stellarator', 'Demo', and 'ITER'.[29] Tokamak is the oldest design invented in 1950s and ITER is latest device. In the Tokamak reactor, atoms of deuterium and tritium are heated in a doughnut-shaped containment till the time their electrons and naked nuclei (called plasma) are separated. Further heating of plasma leads to merger of nuclei and release of free neutrons that act as transporter of heat (generated by fusion) out of plasma and this heat can be used of electricity generation. Currently research is going on in several countries to have a flawless reactor. Scientists are hopeful that within a decade or so, fusion reactors will be ready for commercial electricity generation.

2. *Fuel Cell*

A fuel cell converts chemical energy of a fuel (such as hydrogen) directly into electricity and is similar to a battery[30] with an anode (to supply electrons) and a cathode (to absorb electrons) immersed in an electrolyte (solid or liquid). The difference between the two is that fuel cell can be continuously supplied with fuel and air from an external source unlike a battery that has fixed quantity of fuel material. One of

---

[29]Ref—The Economist Technology Quarterly, December 6th 2014.

[30]https://www.britannica.com/technology/fuel-cell.

the greatest advantages of fuel cells is that they do not emit carbon dioxide or any other air pollutant. Besides, fuel cells do not require recharging (as in batteries) and can continue to supply electric current as long as fuel is supplied. There are different types of fuel cells such as alkali fuel cells that use potassium hydroxide as electrolyte, molten carbonate fuel cells that use sodium or magnesium carbonate as electrolyte, phosphoric acid fuel cells that use phosphoric acid as electrolyte, proton exchange membrane fuel cells with a polymer electrolyte in the form of a thin permeable sheet, and solid oxide fuel cells that use ceramic compounds of metal (calcium and zirconium etc.) oxides as electrolytes.[31]

In a fuel cell using hydrogen as fuel, a catalyst at the anode separates hydrogen molecules into protons and electrons which take different path to the cathode. The electrons pass through external circuit and generate electrical current, whereas protons travel through the electrolyte to the cathode where they combine with oxygen and the electrons to produce water and heat but no carbon dioxide. Fuel cells have found best use in spacecrafts and satellites where energy is required for a very long duration and the electrical current produced by the cell can be directed outside the cell for energizing equipment and illumination, and so on. Globally, there are over 300,000 fuel cells in use and USA generates around 500 megawatts of non-stop power from fuel cells to more than 7500 cars, 40 hydrogen stations, data centers, hospitals, and defense establishments.[32] Germany uses fuel cell vehicles and many countries in Europe and Japan use fuel cells for transportation system.

References

Aayog NITI (2017) Draft national energy policy. National Institution for Transforming India, Government of India, New Delhi http://niti.gov.in/writereaddata/files/new_initiatives/NEP-ID_276

Allen MR, Frame DJ, Huntingford C, Jones CD, Lowe JA, Meinshausen M, Meinshausen N (2009) Warming caused by cumulative carbon emissions towards the trillionth tonne. Nature 458(7242):1163–1166

Bhattacharya SC (2015) Wood energy in India: status and prospects. Energy 85:310–316

Brown LR (2003) Plan B: rescuing a planet under stress and a civilization in trouble. WW Norton & Company

Brown L (2012) World on the edge: how to prevent environmental and economic collapse. Routledge

Brown LR (2013) Eco-economy: building an economy for the earth. Routledge

Chengappa R (2015) Whiz-bang tech for a green tomorrow. How cutting-edge breakthroughs could provide succour by offering an alternative path to cleaner energy and lower emissions. India Today Magazine, 14 Dec 2015

Faostat FAO (2018) Statistical databases. Food and Agriculture Organization of the United Nations

Goodall C (2010) How to live a low-carbon life: the individual's guide to stopping climate change

Government of India (2019) Economic survey 2018–19

International Energy Agency (2018) Key world energy statistics 2018. OECD Publishing

IPCC AR (2007) IPCC fourth assessment report (AR4). IPCC 1:976

---

[31] https://americanhistory.si.edu/fuelcells/basics.htm.

[32] https://www.energy.gov/eere/fuelcells/fuel-cells.

Johnson G (2009) Plugging into the sun. Natl Geographic 216(3):28–53

Kearney AT (2018) Readiness for the future of production report 2018. In: World economic forum

Lok Sabha (2019a) Lok Sabha Question 337 (Archive)

Lok Sabha (2019b) Lok Sabha Question 770 (Archive)

Ministry of Power (2011) Annual report 2010–2011

Musall FD, Onno K (2011) Local acceptance of renewable energy—A case study from southeast Germany. Energy policy 39(6):3252–3260

Nicholson S, Sikina J (eds) (2016) New earth politics: essays from the Anthropocene. MIT Press

NSS (2001) 55th round. National Sample Survey Organisation Ministry of Statistics & Programme Implementation Government of India, New Delhi

Outlook India Energy (2015) World energy outlook special report 2015/international energy agency. URL: https://www.iea.org/publications/freepublications/publication/IndiaEnergyOutl ook_WEO2015.pdf

Planning Commission (2008) Eleventh five year plan 2007–12

Planning Commission (2015) 12th five year plan (2012–17)

Ravindranath NH, Hall DO (1995) Biomass, energy and environment: a developing country perspective from India. Oxford University Press

Razdan PN, Agarwal RK, Singh R (2008) Geothermal energy resources and their potential in India. Earth Sci India 1

Rozenberg J, Davis SJ, Narloch U, Hallegatte S (2015) Climate constraints on the carbon intensity of economic growth. Environ Res Lett 10(9):095006

Schwab KM (2016) The fours industrial revolution [E-source]. KlausMartin Schwab Available at: https://www.foreignaffairs.com/anthologies/2016-01-01/fourth-industrial-revolution

Schwab K (2017) The fourth industrial revolution. Currency

Seibel BA, Fabry VJ (2003) Marine biotic response to elevated carbon dioxide. Adv Appl Biodiversity Sci 4:59–67

Srivastav AK (1992) Strategy for wasteland afforestation in Gujarat. Indian Forester 118(9):623–629

Srivastav A, Srivastav, Nishida (2019) The science and impact of climate change. Springer

Statistics Energy (2018) Central statistics office, national statistical organisation, ministry of statistics and programme implementation, government of India

Submission on Forest Reference Levels for REDD+ submitted by India to UNFCCC on 8 Jan 2018

Xu M, David JM, Kim SH (2018) The fourth industrial revolution: opportunities and challenges. Int J Financial Res 9(2): 90–95

Website: https://www.climate.gov/news-features/understanding-climate/climate-change-atmosp heric-carbon-dioxide

UNFCCC website at—unfccc.int

https://www.un.org/en/sections/issues-depth/climate-change/

https://www.britannica.com/technology/fuel-cell

https://americanhistory.si.edu/fuelcells/basics.htm

https://www.energy.gov/eere/fuelcells/fuel-cells